Undergraduate Texts in Mathematics

Undergraduate Texts in Mathematics

Apostol: Introduction to Analytic
Number Theory.

Armstrong: Basic Topology.

Bak/Newman: Complex Analysis.

Banchoff/Wermer: Linear Algebra
Through Geometry.

Childs: A Concrete Introduction to
Higher Algebra.

Chung: Elementary Probability Theory
with Stochastic Processes.

Croom: Basic Concepts of Algebraic
Topology.

Curtis: Linear Algebra:
An Introductory Approach.

Dixmier: General Topology.

Driver: Why Math?

Ebbinghaus/Flum/Thomas
Mathematical Logic.

Fischer: Intermediate Real Analysis.

Fleming: Functions of Several Variables.
Second edition.

Foulds: Optimization Techniques: An
Introduction.

Foulds: Combination Optimization for
Undergraduates.

Franklin: Methods of Mathematical
Economics.

Halmos: Finite-Dimensional Vector
Spaces. Second edition.

Halmos: Naive Set Theory.

Iooss/Joseph: Elementary Stability and
Bifurcation Theory.

Jänich: Topology.

Kemeny/Snell: Finite Markov Chains.

Klambauer: Aspects of Calculus.

Lang: Undergraduate Analysis.

Lang: A First Course in Calculus. Fifth
Edition.

Lang: Calculus of One Variable. Fifth Edition.

Lang: Introduction to Linear Algebra. Second
Edition.

Lax/Burstein/Lax: Calculus with
Applications and Computing, Volume 1.
Corrected Second Printing.

LeCuyer: College Mathematics with APL.

Lidl/Pilz: Applied Abstract Algebra.

Macki/Strauss: Introduction to Optimal
Control Theory.

Malitz: Introduction to Mathematical
Logic.

Marsden/Weinstein: Calculus I, II, III.
Second edition.

continued after Index

Dennis Stanton   Dennis White

Constructive Combinatorics

With 73 Illustrations

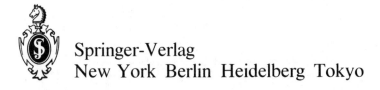

Springer-Verlag
New York Berlin Heidelberg Tokyo

Dennis Stanton
Dennis White
School of Mathematics
University of Minnesota
Minneapolis, MN 55455
U.S.A.

AMS Classifications: 05–01, 05–A05, 05–A15

Library of Congress Cataloging in Publication Data
Stanton, Dennis.
  Constructive combinatorics.
  (Undergraduate texts in mathematics)
  Bibliography: p.
  Includes index.
  1. Combinatorial analysis. I. White, Dennis,
1945– . II. Title. III. Series.
QA164.S79 1986   511'.6   86-6585

Printed and bound by R.R. Donnelley & Sons, Harrisonburg, Virginia.
Printed in the United States of America.

9 8 7 6 5 4 3 2 1

ISBN 0-387-96347-2 Springer-Verlag New York Berlin Heidelberg Tokyo
ISBN 3-540-96347-2 Springer-Verlag Berlin Heidelberg New York Tokyo

Preface

The notes that eventually became this book were written between 1977 and 1985 for the course called Constructive Combinatorics at the University of Minnesota. This is a one-quarter (10 week) course for upper level undergraduate students. The class usually consists of mathematics and computer science majors, with an occasional engineering student. Several graduate students in computer science also attend. At Minnesota, Constructive Combinatorics is the third quarter of a three quarter sequence. The first quarter, Enumerative Combinatorics, is at the level of the texts by Bogart [Bo], Brualdi [Br], Liu [Li] or Tucker [Tu] and is a prerequisite for this course. The second quarter, Graph Theory and Optimization, is not a prerequisite. We assume that the students are familiar with the techniques of enumeration: basic counting principles, generating functions and inclusion/exclusion.

This course evolved from a course on combinatorial algorithms. That course contained a mixture of graph algorithms, optimization and listing algorithms. The computer assignments generally consisted of testing algorithms on examples. While we felt that such material was useful and not without mathematical content, we did not think that the course had a coherent mathematical focus. Furthermore, much of it was being taught, or could have been taught, elsewhere. Graph algorithms and optimization, for instance, were inserted into the graph theory course where they naturally belonged. The computer science department already taught some of the material: the simpler algorithms in a discrete mathematics course; efficiency of algorithms in a more advanced course.

We decided to take as our point of view a decidedly modern trend in combinatorics: the attempt to give algorithmic explanations to combinatorial phenomena. While the systematization of this point of view is modern, its mathematical roots are quite deep, dating back to Euler, Cayley and Sylvester. The resulting course and this book are therefore more mathematically unified and deeper than the course's precursor. Nevertheless, we still believe that the material in this book should be of interest to students in computer science, as well as those in mathematics and other sciences. In fact, this book might provide the jumping-off point for a deeper investigation of related subjects: linear algebra, computational complexity, lattice theory, group theory, representation theory, special functions or mathematical physics.

In this book we use combinatorial algorithms for two purposes. First, a constructive proof of a theorem can be an algorithm. These algorithms often describe a bijection between two finite sets. So we concentrate on interesting mathematical theorems which are proved by bijections. The other purpose is interactive: use the algorithms to investigate interesting mathematical examples. Here the examples are our main focus. An algorithm can be used to generate data related to a problem. It is then up to the students to study these data, formulate as many conjectures as they can, and then prove them. They are not told what the theorems are in advance. Unfortunately, this kind of "research" is usually impossible in most undergraduate mathematics courses.

The material here is more than what can be covered in a 10 week course. Two sections of peripheral interest are §§1.4 and 2.4. Moreover, some of the material in Chapters 3 and 4 (§§3.5-3.7 and §§4.5-4.6) could be considered graduate material. Strictly speaking, each chapter can be presented independently, although we frequently tie together material from different chapters. There are many other topics which would have been suitable for inclusion. One such topic we regretted omitting was the Lagrange inversion formula (see [La] and [Ra]).

The notes are organized in the following way. In Chapter 1 algorithms which list fundamental combinatorial objects are given. They are written in a shorthand version of Pascal (no declaration or i/o statements are given). It is assumed that the students are familiar with a programming language, though not necessarily Pascal. In Chapter 2, a partially ordered set is defined for each object. We concentrate on the Boolean algebra. A number of interesting bijections are given in Chapter 3 for these objects. Finally, we generalize bijections to involutions in Chapter 4. There is some emphasis on tableaux in these last two chapters. Thus they can serve as a combinatorial forerunner to the theory of representations of the symmetric group.

We have included more complete Pascal programs in the Appendix. Furthermore, we would be happy to provide disks (Apple Macintosh Pascal© or Turbo Pascal©) with source code for these programs to interested readers.

The exercises vary from true exercises to very difficult problems. We have assigned each exercise a number from one to four, which we believe is some indication of its difficulty (one is easy, four is hard). Exercises involving a computer are marked with a "C".

Exercises labeled 3C or 4C might be suitable for a term project. We feel strongly that anyone using this book as a text should assign one or more of these exercises. They give the student a chance to use the computer in a non-routine way and to engage in the excitement of mathematical investigation.

We would like to thank the following people who have helped in various ways in the development of this book: David Bressoud, Adriano Garsia, Ira Gessel, Jay Goldman, Jim Joichi, Jeff Remmel, Richard Stanley, Gerard Viennot, Herb Wilf, Gill Williamson and Doron Zeilberger. Finally, we wish to thank the many students who took the courses upon which this material is based. They were our guinea pigs and their feedback has been an important source of direction for us.

This book was prepared using MacWrite©, MacPaint©, MacDraw© and Macintosh Pascal© on our Apple Macintosh© computers, and was printed on an Apple Laserwriter© printer. While the Macintosh gave us a wonderful environment to incorporate text, programs and drawings into a single entity, we cannot say that Apple supports, in an adequate way, mathematical text preparation.

Minneapolis, Minnesota U.S.A.

Dennis W. Stanton
Dennis E. White

Table of Contents

1 Listing Basic Combinatorial Objects   1

1.1   Permutations   2
1.2   Subsets   7
1.3   Integer Partitions   11
1.4   Product Spaces   15
1.5   Set Partitions   18
      Notes   22
      Exercises   22

2 Partially Ordered Sets   26

2.1   Six Posets   26
2.2   Matching the Boolean Algebra   33
2.3   The Littlewood-Offord Problem   39
2.4   Extremal Set Theory   43
      Notes   50
      Exercises   50

3 Bijections   57

3.1   The Catalan Family   59
3.2   The Prüfer Correspondence   64
3.3   Partitions   69
3.4   Permutations   73
3.5   Tableaux   81
3.6   The Schensted Correspondence   85
3.7   Properties of the Schensted Correspondence   93

Notes   102
Exercises   102

4 Involutions   110

4.1   The Euler Pentagonal Number Theorem   111
4.2   Vandermonde's Determinant   114
4.3   The Cayley-Hamilton Theorem   120
4.4   The Matrix-Tree Theorem   125
4.5   Lattice Paths   130
4.6   The Involution Principle   141
      Notes   147
      Exercises   148

Bibliography   156

Appendix   159

A.1   Permutations   159
A.2   Subsets   162
A.3   Set Partitions   164
A.4   Integer Partitions   166
A.5   Product Spaces   167
A.6   Match to First Available   169
A.7   The Schensted Correspondence   171
A.8   The Prüfer Correspondence   176
A.9   The Involution Principle   178

Index   180

CHAPTER 1

Listing Basic Combinatorial Objects

At a basic level, one would expect that constructive combinatorics would address the question of how one constructs the fundamental objects in combinatorics. In fact, "constructing" these objects could mean providing an algorithm for listing all of them, or it could mean generating one of them at random. While both questions are of interest, we shall concentrate on the first.

It is frequently useful in combinatorics to have such listing algorithms. The most obvious application is to use the algorithms to produce computer programs which test conjectures and theorems for combinatorial objects. Better yet, conjectures might be discovered in the resulting data. In another direction, the algorithm itself could be of mathematical interest. The algorithm might be a *proof* of a theorem. For instance, the existence of an algorithm which lists permutations by transposing adjacent objects *proves* that any permutation can be written as a product of adjacent transpositions.

Any such list of objects gives the objects a *linear order*. This means that if a and b are objects, then we can say that $a < b$ if a precedes b on the list. This linear order clearly gives a ranking function on the objects. One might expect that the first object has rank one, and so on. However, we shall find it more useful to define the *Rank* of an object as the number of objects in the list which precede it. So if a is first, $Rank(a) = 0$. If the list has N objects, the function Rank must map these objects to the set $\{0, 1, \ldots, N-1\}$.

While the theoretical definition of Rank is obvious, it is often not at all clear how to construct Rank without listing all of the objects. In fact, algorithms to rank objects (find Rank(a)) and unrank integers (find $Unrank(i) = Rank^{-1}(i)$, $i \in \{0, 1, \ldots, N-1\}$) often give great insight into the algorithm.

In this chapter we give listing algorithms for these combinatorial objects: permutations, subsets of a set, integer partitions, set partitions and product spaces. Each algorithm will be based upon a recursive formula for the number of objects listed. For example, subsets may be listed by using the combinatorial interpretation of Pascal's triangle.

The most important ranking function will use "lexicographic" ordering. It can be used for virtually any combinatorial object. We shall see in Chapter 2 that it also has many remarkable and surprising theoretical properties.

§1.1 Permutations

A *permutation* of n distinct objects of length k is an ordered arrangement of any k of the objects. For instance, the permutations of {a, b, c, d} of length two are ab, ac, ad, ba, bc, bd, ca, cb, cd, da, db and dc. The next proposition is clear.

PROPOSITION 1.1 *The number of permutations of* n *objects of length* k *is* $n(n-1)\cdots(n-k+1)$.

Sometimes we shall write $(n)_k$ (called the *falling factorial*) for $n(n-1)\cdots(n-k+1)$.

A permutation of n objects of length n is frequently called a permutation of n objects (or simply a *permutation of n*). It is clear that we can take the set $[n] = \{1, 2, \ldots, n\}$ for the n objects. We shall frequently use this notation. Proposition 1.1 shows that the number of permutations of n is $(n)_n = n!$.

Perhaps the most natural ordering of the permutations of n is *lexicographic (lex) order*. We say that π precedes σ in lex order, if, for some i, the first i entries of π and σ are the same, and the (i+1)th entry of π is less than the (i+1)th entry of σ. The lex list of the permutations of 3 is 123, 132, 213, 231, 312 and 321. This ordering is quite simple. You are asked to consider it in Exercises 2 and 3. We shall return to lex order in §1.2.

We consider instead an algorithm to list all permutations of n that is due to Johnson [Joh] and Trotter [T]. It is based on a "combinatorial proof" of $n! = n(n-1)!$: for each of the (n-1)! permutations of [n-1], there are n "positions" into which n may be inserted. The algorithm has the property that each permutation differs from its predecessor by only a transposition of adjacent symbols. The lex list does not have this property.

How does the algorithm work? Suppose we have the list for permutations of [n-1]: $\pi^{(0)}$, $\pi^{(1)}$, ... Then we construct the list for permutations of [n] by inserting n into each of the n possible positions of each $\pi^{(i)}$. The insertions go from left to right if i is odd and right to left if i is even. The lists for n = 1, 2, 3 and 4 are given below with the recursive structure indicated.

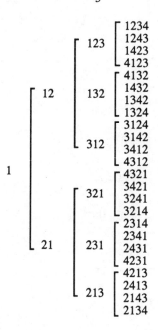

The Johnson-Trotter algorithm is listed in Algorithm 1. The permutations π are given by $\pi = (\pi[1], \ldots, \pi[n])$. (Usually we will write π as $(\pi_1, \ldots, \pi_n)$ but in algorithms we will sometimes use [ ] instead of subscripts.) The inverse of the permutation π is also used, so $\pi[\pi^{-1}[m]] = m$. In order to keep track of which digits are moving, we use the set of active digits, A. Initially all digits > 1 are active. If a digit i comes to a boundary, it becomes passive (is deleted from A), and all digits $>$ i become active. The directions that the digits are moving are given by the direction vector $(d[1], \ldots, d[n])$; $d[i] = 1$ is to the right and $d[i] = -1$ is to the left. Initially, each $d[i]$ is -1 and $(\pi[1], \ldots, \pi[n]) = (1, 2, \ldots, n)$. The variables $\pi[0]$ and $\pi[n+1]$ are used for "the boundary positions" and are fixed at n+1. The algorithm ends when A is empty.

ALGORITHM 1: *Permutation List*

begin
      for $i \leftarrow 1$ **to** $n + 1$ **do**
            $\pi[i] \leftarrow i$
            $\pi^{-1}[i] \leftarrow i$
            $d[i] \leftarrow -1$
      $\pi[0] \leftarrow n + 1$
      $A \leftarrow \{2, \ldots, n\}$

```
        Done ← false
        while not Done do
            Print(π)
            if A ≠ ∅ then
                m ← max{i : i ∈ A}
                j ← π⁻¹[m]
                π[j] ← π[j + d[m]]
                π[j + d[m]] ← m
                π⁻¹[m] ← π⁻¹[m] + d[m]
                π⁻¹[π[j]] ← j
                if m < π[j + 2·d[m]] then
                    d[m] ← − d[m]
                    A ← A − {m}
                A ← A ∪ {m + 1, ... , n}
            else
                Done ← true
    end.
```

The proof that Algorithm 1 works is inductive. It is clear that it will cause n to sweep back and forth across the permutation, and, at each boundary, construct a new permutation of [n−1]. As n sweeps across, no d[i] changes. Thus the only changes in (d[1], ... , d[n−1]) occur when n reaches a boundary. These are precisely the changes that would be encountered in the permutation list for [n−1].

The function Rank for Algorithm 1 can now be given. Clearly, Rank(1, 2, ... , n) = 0. Let $\pi = (\pi[1], ... , \pi[n])$ be a permutation of n, and let π' be the permutation of [n−1] which is π with n deleted. Note that n has made Rank(π') complete sweeps. Its last incomplete sweep is right to left if Rank(π') is even and left to right if Rank(π') is odd. If n occupies position j in π (that is, $\pi[j] = n$), we find

$$(1.1) \qquad \text{Rank}(\pi) = n\,\text{Rank}(\pi') + \begin{cases} j - 1 & \text{if Rank}(\pi') \text{ is odd} \\ n - j & \text{if Rank}(\pi') \text{ is even.} \end{cases}$$

For example, suppose $\pi = (5, 1, 6, 2, 3, 7, 4)$. Then (1), (1, 2), (1, 2, 3) and (1, 2, 3, 4) all have rank 0. So

Rank(5, 1, 2, 3, 4) = 5·0 + (5 − 1) = 4 since 0 is even,
Rank(5, 1, 6, 2, 3, 4) = 6·4 + (6 − 3) = 27 since 4 is even, and

$$\text{Rank}(5, 1, 6, 2, 3, 7, 4) = 7\cdot 27 + (6 - 1) = 194.$$

ALGORITHM 2: *Rank Permutation*

begin

    $R \leftarrow 0$

    for $i \leftarrow 1$ **to** n **do**

        $\text{Moves} \leftarrow |\{j : j < i \text{ and } \pi^{-1}[j] < \pi^{-1}[i]\}|$

        if R **odd then**

            $\text{remainder} \leftarrow \text{Moves}$

        else

            $\text{remainder} \leftarrow i - 1 - \text{Moves}$

        $R \leftarrow i\cdot R + \text{remainder}$

    $\text{Rank}(\pi) \leftarrow R$

end.

To unrank M, the algorithm is reversed.

ALGORITHM 3: *Unrank Permutation*

begin

    for $j \leftarrow 1$ **to** n **do**

        $\pi[j] \leftarrow 0$

    $P \leftarrow M;$

    for $j \leftarrow n$ **downto** 1 **do**

        $R \leftarrow P \bmod j$

        $P \leftarrow \lfloor P/j \rfloor$

        if P **odd then**

            $k \leftarrow 0$

            $\text{Dir} \leftarrow 1$

        else

            $k \leftarrow n + 1$

            $\text{Dir} \leftarrow -1$

        $C \leftarrow 0$

        repeat

            $k \leftarrow k + \text{Dir}$

            if $\pi[k] = 0$ **then**

                $C \leftarrow C + 1$

        until $C = R + 1$

        $\pi[k] \leftarrow j$

Unrank(M) ← π

end.

In fact, Algorithms 2 and 3 have a theoretical consequence.

THEOREM 1.2 *Every integer* k *satisfying* $0 \leq k \leq n! - 1$ *can be uniquely represented as*

$$k = \sum_{i=1}^{n} (n)_{n-i} \, b_i$$

where b_i *satisfies* $0 \leq b_i \leq i - 1$.

Proof Algorithms 2 and 3 establish a one-to-one correspondence between permutations of n and sequences $(b_1, \dots, b_n)$ where $0 \leq b_i < i$. Iterating the recurrence relation (1.1) for Rank(π) gives the theorem.

Permutations of n can be thought of as one-to-one functions from the set [n] = $\{1, 2, \dots, n\}$ to itself. Recall that the *product* of two permutations of n is their composition as functions on [n]. So if $\pi = (5, 4, 1, 3, 2)$ and $\mu = (3, 5, 1, 4, 2)$, then $\pi \circ \mu = (1, 2, 5, 3, 4)$ because

$$
\begin{array}{ccccc}
1 & \xrightarrow{\mu} & 3 & \xrightarrow{\pi} & 1 \\
2 & \xrightarrow{\mu} & 5 & \xrightarrow{\pi} & 2 \\
3 & \xrightarrow{\mu} & 1 & \xrightarrow{\pi} & 5 \\
4 & \xrightarrow{\mu} & 4 & \xrightarrow{\pi} & 3 \\
5 & \xrightarrow{\mu} & 2 & \xrightarrow{\pi} & 4 .
\end{array}
$$

Any permutation of [n] can be written as a product of disjoint cycles in cycle notation. For $\pi = (3, 5, 1, 4, 2)$, we have $\pi = (1\ 3)(2\ 5)(4)$. Note that any 1-cycle (a cycle of length one in π) corresponds to a fixed point of π. In the example $\pi(4) = 4$ is a fixed point. A *transposition* is a permutation with exactly two points that are not fixed. Thus a transposition fixes all but two points, say j and k, which form a two cycle (j k) of π. An *adjacent transposition* is a transposition of the form (j j+1). We now note that Algorithm 1 also has a theoretical consequence.

THEOREM 1.3 *Every permutation can be written as a product of adjacent transpositions.*

Proof This follows from Algorithm 1, in which each permutation is obtained from its predecessor by an adjacent transposition.

So far we have used two representations of permutations: *one-line notation* $\pi = (3, 5, 1, 4, 2)$ and *cycle notation* $\pi = (1\ 3)(2\ 5)(4)$. Another notation is *two-line notation*

$$\begin{pmatrix} 1\,2\,3\,4\,5 \\ 3\,5\,1\,4\,2 \end{pmatrix}$$

which lists π_i under i. There are several other ways of identifying permutations. One example, which will be considered in Chapter 2, is the *inversion sequence* of π. Let $\pi = (\pi_1, \ldots, \pi_n)$. The inversion sequence of π, $(a_1, \ldots, a_n)$, is defined by $a_i = |\{j : j < i \text{ and } \pi_j > \pi_i\}|$. In the example above, the inversion sequence is $(0, 0, 2, 1, 3)$. Clearly $0 \le a_i \le i - 1$. In Exercise 7 you are asked to reconstruct π from its inversion sequence. We shall return to this statistic on permutations in Chapters 2 and 3.

§1.2 Subsets

One of the fundamental building blocks of combinatorics is the binomial coefficient. Recall that the number of k-element subsets of the set $[n] = \{1, 2, \ldots, n\}$ is the binomial coefficient

$$\binom{n}{k} = \frac{n!}{k!\,(n-k)!}.$$

Recall also that the binomial theorem is

$$(2.1) \qquad (x+y)^n = \sum_{k=0}^{n} \binom{n}{k} x^k y^{n-k}$$

where n is a non-negative integer and x and y are complex numbers.

There are many formulas relating binomial coefficients. Frequently these formulas have combinatorial interpretations as *bijections*. A bijection between two sets

A and B is a function $f : A \rightarrow B$ which is one-to-one and onto. Clearly if there is a bijection between two finite sets, those two sets contain the same number of elements. In fact, we have already seen a bijection: the Rank function in §1.1. The symmetry relation

$$(2.2) \qquad \binom{n}{k} = \binom{n}{n-k}$$

can easily be shown by a bijection. Let A be the set of k-element subsets of [n] and let B be the set of (n–k)-element subsets of [n]. Complementation is a bijection from A to B, and (2.2) is established.

In this section we shall use the Pascal triangle property of the binomial coefficients to list all k-element subsets of [n]. It is

$$(2.3) \qquad \binom{n}{k} = \binom{n-1}{k-1} + \binom{n-1}{k}.$$

A bijective proof of (2.3) is easy. Split the set of k-subsets of [n] into two sets: those containing 1 and those not containing 1.

A k-element subset of [n] is a k-tuple $(v_1, v_2, \ldots, v_k)$ where $v_1 < v_2 < \ldots < v_k$. Our listing uses *lexicographic* (dictionary) order on such k-tuples. We say that $(v_1, v_2, \ldots, v_k)$ is before $(w_1, w_2, \ldots, w_k)$ in lexicographic order if, for some $1 \le j \le k - 1$, $(v_1, v_2, \ldots, v_j) = (w_1, w_2, \ldots, w_j)$ and $v_{j+1} < w_{j+1}$. This is the same ordering that words with distinct letters have in a dictionary. So the 3-element subsets of [6] in lex order are 123, 124, 125, 126, 134, 135, 136, 145, 146, 156, 234, 235, 236, 245, 246, 256, 345, 346, 356 and 456. The first 10 subsets on this list begin with a 1. They are listed according to lex order of the two element subsets of $\{2, 3, 4, 5, 6\}$. The remaining sets are listed in lex order for the three-element subsets of $\{2, 3, 4, 5, 6\}$. Thus the lex list is given exactly by the combinatorial proof of (2.3).

There is another order closely related to lex order, called *colexicographic* (colex) order. This order is defined by reading the k-tuple $(v_1, v_2, \ldots, v_k)$ from right to left instead of left to right. The colex list for 3-element subsets of [6] is 123, 124, 134, 234, 125, 135, 235, 145, 245, 345, 126, 136, 236, 146, 246, 346, 156, 256, 356 and 456. It is clear that this list is just the colex list for 3-element subsets of [5], followed by the colex list of 2-element subsets of [5], with 6 adjoined. Again it is a version of (2.3). In fact, the lex list can be found from the colex list (see Exercise 10).

The algorithm for the colex list is based upon a successor algorithm. The subset which follows $(v_1, v_2, \ldots, v_k)$ in colex order is obtained by finding the smallest j such that $v_j + 1 < v_{j+1}$ (or $j = k$ if there is no such j), replacing v_j by $v_j + 1$, and replacing $(v_1, v_2, \ldots, v_{j-1})$ by $(1, 2, \ldots, j-1)$.

ALGORITHM 4: *Subset List*

begin

    for $i \leftarrow 1$ **to** k **do**

        $v_i \leftarrow i$

    $v_{k+1} \leftarrow n + 1$

    Done $\leftarrow$ **false**

    while not Done **do**

        Print(v)

        if $v_1 < n - k + 1$ **then**

            $j \leftarrow 0$

            repeat

                $j \leftarrow j + 1$

            until $v_{j+1} > v_j + 1$

            $v_j \leftarrow v_j + 1$

            for $i \leftarrow 1$ **to** $j - 1$ **do**

                $v_i \leftarrow i$

        else

            Done $\leftarrow$ **true**

end.

To find $\text{Rank}(v_1, v_2, \ldots, v_k)$, note that $(v_1, v_2, \ldots, v_k)$ is preceded by all k-element subsets of $[v_k - 1]$, and by all k-element subsets $(\tilde{v}_1, \tilde{v}_2, \ldots, \tilde{v}_k)$, where $(\tilde{v}_1, \tilde{v}_2, \ldots, \tilde{v}_{k-1})$ precedes $(v_1, v_2, \ldots, v_{k-1})$ in the colex list of $(k-1)$-element subsets and $\tilde{v}_k = v_k$. This is

$$(2.4) \qquad \text{Rank}(v_1, \ldots, v_k) = \binom{v_k - 1}{k} + \text{Rank}(v_1, \ldots, v_{k-1}).$$

For example, if $(v_1, v_2, \ldots, v_k) = (3, 4, 7) \subset [7]$, $\text{Rank}(3, 4, 7) = 25$ since it is preceded by

$$
\begin{aligned}
&\left.\begin{matrix} 123 \\ 456 \end{matrix}\right\} \quad \text{3-element subsets of [6]} \\[6pt]
&\left.\begin{matrix} 127 \\ 237 \end{matrix}\right\} \quad \text{2-element subsets of [3]} \\[6pt]
&\left.\begin{matrix} 147 \\ 247 \end{matrix}\right\} \quad \text{1-element subsets of [2]}
\end{aligned}
$$

and

$$
\binom{6}{3} + \binom{3}{2} + \binom{2}{1} = 25.
$$

ALGORITHM 5: *Rank Subset*

begin

  $R \leftarrow 0$

  for $i \leftarrow 1$ **to k do**

   $R \leftarrow R + \binom{v_i - 1}{i}$

  $\text{Rank}(v) \leftarrow R$

end.

Algorithm 5 shows that Rank is a bijection from $(v_1, v_2, \ldots, v_k)$ with $1 \le v_1 < v_2 < \ldots < v_k \le n$ to the set $[N-1] \cup \{0\}$,

$$
N = \binom{n}{k}
$$

so that we have proved the following theorem.

THEOREM 2.1 *Any integer* m *satisfying*

$$
0 \le m \le \binom{n}{k} - 1
$$

can be uniquely expressed as

$$
m = \sum_{i=1}^{k} \binom{v_i - 1}{i}
$$

for some positive integers satisfying $1 \le v_1 < v_2 < \ldots < v_k \le n$.

To find Unrank(m) for a given k and find the largest binomial coefficient $\binom{i}{k}$ that is $\leq m$. Then put $v_k = i + 1$, subtract that binomial coefficient from m, and repeat with k replaced by $k - 1$.

ALGORITHM 6: *Unrank Subset*

begin

      $R \leftarrow m$

      for $i \leftarrow k$ **downto** 1 **do**

          $p \leftarrow i - 1$

          repeat

              $p \leftarrow p + 1$

          until $\binom{p}{i} > R$

          $R \leftarrow R - \binom{p-1}{i}$

          $v_i \leftarrow p$

      Unrank(m) $\leftarrow v$

end.

Note that the condition $1 \leq v_1 < v_2 < \ldots < v_k \leq n$ is not immediately obvious from Algorithm 6. We know that it holds because Unrank is the inverse function of Rank. You are asked to verify this condition in Exercise 11.

§1.3 Integer Partitions

The third combinatorial object we consider is a partition of an integer. A k-tuple of positive integers $\lambda = (\lambda_1, \ldots, \lambda_k)$ is an *integer partition of* n if $\lambda_1 + \lambda_2 + \ldots + \lambda_k = n$ and $\lambda_1 \geq \lambda_2 \geq \ldots \geq \lambda_k \geq 1$. The *number of parts* of λ is k. An example of a partition of 12 into 6 parts is $\lambda = (4, 2, 2, 2, 1, 1)$. Alternatively, we can completely describe λ by giving the number of times that a part i occurs, called the *multiplicity* of i. In this notation $\lambda = 4^1 \, 2^3 \, 1^2$, because λ has one 4, three 2's and two 1's.

A useful way of picturing a partition is an array of squares, or cells, left justified, in decreasing order. For example, $4^1 \, 2^3 \, 1^2$ is given by

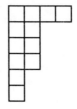

Such diagrams are called *Young diagrams* or *Ferrers diagrams*.

The *conjugate partition* of λ, denoted λ', is the partition obtained from λ by interchanging the rows and columns of the Ferrers diagram of λ. In other words, just transpose the Ferrers diagram of λ. For $\lambda = 4^1\, 2^3\, 1^2$, $\lambda' = 6^1\, 4^1\, 1^2$. It is clear that the map conj: $\lambda \to \lambda'$ gives the following theorem.

THEOREM 3.1 *The number of partitions of* n *with* k *parts is equal to the number of partitions of* n *whose largest part is* k.

Proof The map conj takes the first column of λ to the first row of λ'. So if λ has k parts, the largest part of λ' is k. Because two applications of conj yields the identity map, conj is a bijection.

The listing algorithm for partitions uses *reverse lex* order. Given $\lambda = (\lambda_1, \ldots, \lambda_k)$ and $\mu = (\mu_1, \mu_2, \ldots, \mu_m)$, we say λ precedes μ if, upon reading left to right, the first entry i for which $\lambda_i \neq \mu_i$, satisfies $\lambda_i > \mu_i$. Thus, the reverse lex list for $n = 7$ is

$$7$$
$$6\,1$$
$$5\,2$$
$$5\,1^2$$
$$4\,3$$
$$4\,2\,1$$
$$4\,1^3$$
$$3^2\,1$$
$$3\,2^2$$
$$3\,2\,1^2$$
$$3\,1^4$$

$$2^3 \, 1$$
$$2^2 \, 1^3$$
$$2 \, 1^5$$
$$1^7.$$

Algorithm 7 uses the multiplicities $(m_1, m_2, \ldots, m_k)$ of the parts $(p_1, p_2, \ldots, p_k)$ to describe the partition λ. We require that $m_i \neq 0$ and that $p_1 > p_2 > \ldots > p_k$. Thus, the vectors $(m_1, m_2, m_3, m_4) = (2, 1, 4, 1)$ and $(p_1, p_2, p_3, p_4) = (6, 3, 2, 1)$ correspond to $6^2 \, 3^1 \, 2^4 \, 1^1$, a partition of 24. The number of parts of the partition is kept in the variable r.

Given λ, what is its successor? Find the smallest part p of λ which is not a 1. Then use p and all of the 1's to create as many parts of size $(p-1)$ as possible. The "leftovers" become one part. For $5 \, 4^2 \, 1^4$, $p = 4$ and we change $4 \, 1^4$ to $3^2 \, 2$. So $5 \, 4 \, 3^2 \, 2$ follows $5 \, 4^2 \, 1^4$.

ALGORITHM 7: *Integer Partition List*

begin

    $p_1 \leftarrow n$

    $m_1 \leftarrow 1$

    $r \leftarrow 1$

    Done $\leftarrow$ **false**

    while not Done **do**

        Print(λ)

        if $p_r > 1$ **or** $r > 1$ **then**

            if $p_r = 1$ **then**

                $s \leftarrow p_{r-1} + m_r$

                $k \leftarrow r - 1$

            else

                $s \leftarrow p_r$

                $k \leftarrow r$

        $w \leftarrow p_k - 1$

        $u \leftarrow \lfloor s/w \rfloor$

        $v \leftarrow s \bmod w$

        $m_k \leftarrow m_k - 1$

        if $m_k = 0$ **then**

            $k1 \leftarrow k$

14

```
else
        k1 ← k + 1
    m_k1 ← u
    p_k1 ← w
    if v = 0 then
            r ← k1
    else
            m_{k1+1} ← 1
            p_{k1+1} ← v
            r ← k1 + 1
else
        Done ← true
end.
```

One can ask how to construct the Rank and Unrank functions and if they give explicit representation theorems, such as Theorems 1.2 and 2.1. Since the list is reverse lex, it might make sense to count the partitions which <u>follow</u> the given partition and subtract that number from $p(n)$, the total number of partitions of n. Let $S(\lambda)$ denote the number of partitions which follow λ. Certainly all partitions of n whose largest part is less than λ_1 will follow λ. Let $p(n, k)$ denote the number of partitions of n whose largest part is $< k$. Then

(3.1) $S(\lambda) = p(n, \lambda_1) + S(\lambda^*)$

where

(3.2) $\lambda^* = (\lambda_2, \dots, \lambda_k).$

Iterating (3.1) gives the following theorem.

THEOREM 3.2 *Any integer* i *satisfying* $0 \le i \le p(n) - 1$ *can be uniquely represented by*

$$i = \sum_{j=1}^{k} p(n_j, \lambda_j)$$

for some sequence of positive integers $n \ge \lambda_1 \ge \lambda_2 \ge \dots \ge \lambda_k \ge 1$, *where* $n_j = n - (\lambda_1 + \lambda_2 + \dots + \lambda_{j-1})$, $n_1 = n$.

Unfortunately, there are no simple closed formulas for $p(n)$ or $p(n, k)$. However, the $p(n, k)$ can be computed via a recursion, as can $p(n)$. In Exercises 13 and 14 we ask you to give these recursions and the Rank and Unrank algorithms.

In §4.1 we shall give a more efficient method for computing $p(n)$. In §3.3 we give bijective proofs of several partition theorems. There are also a number of exercises in this chapter and Chapter 3 concerning partitions.

§1.4 Product Spaces

Suppose we need a list of all of the subsets of $[n]$, not just the k-element subsets. We could use Algorithm 4 for $k = 0, 1, \dots, n$ and paste the lists together. In this section we shall give a more natural listing using product spaces. The number of subsets of $[n]$ is 2^n, which is the size of the n-fold Cartesian product $\{0, 1\}^n = \{0, 1\} \times \dots \times \{0, 1\}$. The reason is clear. There is a bijection between all subsets $A \subset [n]$ and all n-tuples of 0's and 1's. For example, $\{3, 4, 6, 8\} \subset [9]$ corresponds to the 9-tuple $(0, 0, 1, 1, 0, 1, 0, 1, 0)$.

A *Gray code* G is a list of the elements of $\{0, 1\}^n$ such that two adjacent n-tuples differ in exactly one component, including the first and last n-tuples. Another description of a Gray code is a Hamiltonian cycle in the n-dimensional cube. From the bijection, a Gray code is equivalent to a list of the subsets of $[n]$, where two adjacent subsets differ by only one element. For $n = 3$ an example of such a code is

$$(0, 0, 0)$$
$$(0, 0, 1)$$
$$(0, 1, 1)$$
$$(0, 1, 0)$$
$$(1, 1, 0)$$
$$(1, 1, 1)$$
$$(1, 0, 1)$$
$$(1, 0, 0).$$

A Gray code can be constructed inductively. Suppose $G(n)$ is one such code for $[n]$. To construct $G(n+1)$, insert a 0 before each of the n-tuples on the list $G(n)$. Follow this list by $G(n)$ backwards, with 1 inserted before each n-tuple. The example above is $G(3)$ with the initial condition $G(1) = 0, 1$.

More generally, we shall give an algorithm for the n-tuples in $\{0, 1, \dots, m_1-1\} \times \{0, 1, \dots, m_2-1\} \times \dots \times \{0, 1, \dots, m_n-1\}$, for positive integers $(m_1, m_2, \dots, m_n)$. Again two adjacent n-tuples will differ in exactly one component.

The main idea for Algorithm 8 is the same as Algorithm 1. This time component i of the n-tuple (v_i) will increase and decrease from 0 to $m_i - 1$. We use the set A to indicate which components are changing. The "direction" vector $(d_1, d_2, \ldots, d_n)$ indicates whether a component is increasing or decreasing. The largest active component is used, just as in Algorithm 1 the largest active digit is used. The set B keeps track of those i such that $m_i = 1$. This set is made permanently passive.

ALGORITHM 8: *Product Space List*

begin

    for $i \leftarrow 1$ **to** n **do**

        $v_i \leftarrow 0$

        $d_i \leftarrow 1$

    $A \leftarrow B^c$

    Done $\leftarrow$ **false**

    while not Done **do**

        Print(v)

        if $A \neq \varnothing$ **then**

            Done $\leftarrow$ **false**

            $p \leftarrow \max\{i : i \in A\}$

            $v_p \leftarrow v_p + d_p$

            if $v_p = m_p - 1$ **or** $v_p = 0$ **then**

                $d_p \leftarrow -d_p$

                $A \leftarrow A - \{p\}$

                $A \leftarrow A \cup (\{p+1, \ldots, n\} \cap B)$

        else

            Done $\leftarrow$ **true**

end.

It is clear that Algorithm 8 will list all $(0, v_2, \ldots, v_n)$ and then change $v_1 = 0$ to $v_1 = 1$ when $A = \{1\}$. The next part of the list (for $v_1 = 1$) will be the $v_1 = 0$ list backwards. The algorithm continues in this way. For $m_1 = m_2 = \ldots = m_n = 2$, Algorithm 8 gives the Gray code $G(n)$.

There is an unexpected bonus from Algorithm 8. If we take $m_i = i$, the product space will have n! elements. This list is precisely the inversion sequence list of the permutations in Algorithm 1.

The Rank algorithm is very similar to Algorithm 2. Let

(4.1) $r_n = Rank(v_1, \ldots, v_n)$.

This time the "clumps" have size m_n, from the last component v_n ranging in $0 \le v_n \le m_n - 1$. So $(v_1, \ldots, v_n)$ has $m_n \cdot r_{n-1}$ n-tuples in clumps preceding it. The direction d_n for the clump of $(v_1, \ldots, v_n)$ depends upon the parity of r_{n-1}. We have

(4.2) $r_n = m_n r_{n-1} + \begin{cases} m_n - v_n - 1 & \text{if } r_{n-1} \text{ is odd} \\ v_n & \text{if } r_{n-1} \text{ is even.} \end{cases}$

ALGORITHM 9: *Rank Product Space*

begin

    $R \leftarrow 0$

    for $i \leftarrow 1$ **to** n **do**

        if R **odd then**

            $C \leftarrow m_i - v_i - 1$

        else

            $C \leftarrow v_i$

        $R \leftarrow m_i \cdot R + C$

    $Rank(v) \leftarrow R$

end

ALGORITHM 10: *Unrank Product Space*

begin

    $S \leftarrow R$

    for $i \leftarrow n$ **downto** 1 **do**

        $v_i \leftarrow S \bmod m_i$

        $S \leftarrow \lfloor S/m_i \rfloor$

        if S **odd then**

            $v_i \leftarrow m_i - v_i - 1$

    $Unrank(R) \leftarrow v$

end.

§1.5 Set Partitions

A *set partition* is analogous to an integer partition. Instead of writing the integer n as a sum of positive integers, we write the set [n] as a disjoint union of subsets. These disjoint subsets of [n] are called *blocks* of the set partition. There are 15 set partitions of [4]. They are

1234	12-34	14-2-3
123-4	13-24	23-1-4
124-3	14-23	24-1-3
134-2	12-3-4	34-1-2
234-1	13-2-4	1-2-3-4 .

The number of set partitions of [n] is called the *Bell number* B_n. The number of set partitions of [n] with k blocks is called the *Stirling number of the second kind* $S(n, k)$. (We shall discuss the Stirling number of the first kind in §3.4.) So we see that $B_4 = 15$ and $S(4, 1) = 1$, $S(4, 2) = 7$, $S(4, 3) = 6$ and $S(4, 4) = 1$.

There is an analogue to Pascal's triangle (2.3) for the Stirling numbers of the second kind. It is

(5.1) $S(n, k) = S(n - 1, k - 1) + k \cdot S(n - 1, k)$.

The bijective proof of (5.1) is easy. If $n \in [n]$ lies in a block by itself, the remaining blocks form a set partition of [n−1]. Otherwise n lies in one of the k blocks of a set partition of [n−1].

We shall use another bijection to list all set partitions of [n]. A *restricted growth function* on [n] (or RG function) is a vector $(v_1, v_2, \ldots, v_n)$ satisfying $v_1 = 1$ and $v_i \le \max\{v_1, \ldots, v_{i-1}\} + 1$. The 15 RG functions on [4] are

1111	1211	1223
1112	1212	1231
1121	1213	1232
1122	1221	1233
1123	1222	1234.

THEOREM 5.1 *There is a bijection between RG functions on* [n] *and set partitions of* [n].

Proof Let A be the set of all RG functions on [n], and let B be the set of all set partitions of [n]. We define a map $\Phi : A \to B$ which is the required bijection. Put

$\Phi((v_1, v_2, \ldots, v_n)) = S_1 \cup S_2 \cup \ldots \cup S_k$, where $S_i = \{j : v_j = i\}$. The map Φ certainly maps A to B. We must verify that it is one-to-one and onto. To do this, we explicitly construct the inverse function to Φ, $\bar{\Phi} : B \to A$.

Let $S_1 \cup S_2 \cup \ldots \cup S_k \in B$ be a set partition. Assume that the blocks have been ordered in the following way: $1 \in S_1$, $\min\{i \in [n] - S_1\} \in S_2$ and $\min\{i \in [n] - S_1 \cup S_2 \cup \ldots \cup S_{j-1}\} \in S_j$. This just means that we order the blocks so that block S_j contains the smallest element not in preceding blocks. We define $\bar{\Phi}(S_1 \cup S_2 \cup \ldots \cup S_k) = (v_1, v_2, \ldots, v_n)$ by $v_i = j$ if and only if $i \in S_j$. We must check that $(v_1, v_2, \ldots, v_n)$ is an RG function. Clearly $v_1 = 1$. Let $i \in [n]$ and put $m = \max\{v_1, v_2, \ldots, v_{i-1}\}$. Then $[i-1] \subset S_1 \cup S_2 \cup \ldots \cup S_m$ and either $i \in S_1 \cup S_2 \cup \ldots \cup S_m$ or i is the smallest number outside this union. In the first case $v_i \leq m$, while in the second case i is in the block S_{m+1}, which means that $v_i = m + 1$. This verifies that $v_i \leq \max\{v_1, \ldots, v_{i-1}\} + 1$.

It is now easy to verify that $\Phi \circ \bar{\Phi}$ and $\bar{\Phi} \circ \Phi$ are identity maps on B and A respectively. This implies that Φ is a bijection.

Algorithm 11 lists all RG functions on $[n]$ in lex order. We begin with $11 \ldots 1$ and always increase an entry by one. We find which entries can be increased by one and still retain the RG condition. So we record the "legal" maximum $m_i = \max\{v_1, \ldots, v_i\} + 1$.

ALGORITHM 11: *Restricted Growth Function List*

begin
>   **for** $i \leftarrow 1$ **to** n **do**
>> $v_i \leftarrow 1$
>> $m_i \leftarrow 2$
>   Done $\leftarrow$ **false**
>   **while not** Done **do**
>>   Print(v)
>> $j \leftarrow n + 1$
>>   **repeat**
>>> $j \leftarrow j - 1$
>>   **until** $v_j \neq m_j$
>>   **if** $j > 1$ **then**

$$v_j \leftarrow v_j + 1$$

for $i \leftarrow j + 1$ **to** n **do**

$$v_i \leftarrow 1$$

if $v_j = m_j$ **then**

$$m_i \leftarrow m_j + 1$$

else

$$m_i \leftarrow m_j$$

else

Done $\leftarrow$ **true**

end.

The Rank and Unrank algorithms are more difficult. We need the number of ways of "finishing" an RG function with a specified beginning. This is because the number of predecessors of $(v_1, v_2, \ldots , v_n)$ is

$$\sum_{i=1}^{n} \sum_{j=1}^{v_i - 1} A_{ij}(v_1, \ldots , v_n)$$

where $A_{ij}(v_1, v_2, \ldots , v_n)$ is the number of RG functions which begin $(v_1, v_2, \ldots , v_{i-1}, j)$. Two observations can be made about $A_{ij}(v_1, v_2, \ldots , v_n)$. First, it depends only upon i and the largest value in $\{v_1, v_2, \ldots , v_{i-1}, j\}$. Second, $j \leq \max\{v_1, v_2, \ldots , v_{i-1}\}$ by the RG condition on v_i, so $\max\{v_1, v_2, \ldots , v_{i-1}, j\}$ $= \max\{v_1, v_2, \ldots , v_{i-1}\} = u_i$. Thus, we may rewrite the above expression as

$$\sum_{i=1}^{n} d_{n-i, u_i}(v_i - 1)$$

where $d_{m,t}$ is the number of ways of "finishing" the last m positions if t is the maximum of the first $n - m$ positions.

Finally, note that

(5.2) $$d_{m,t} = t\, d_{m-1,t} + d_{m-1,t+1}$$

because we may place either $t + 1$ in the $n - m + 1$ position, leaving $m - 1$ positions to fill, with largest value now $t + 1$; or we may place $1, 2, \ldots , t$ in the $n - m + 1$ position, leaving $m - 1$ positions to fill, with largest value still t.

ALGORITHM 12: *Rank Restricted Growth Function*

begin

    $u_1 \leftarrow 1$

    for $i \leftarrow 2$ **to** n **do**

        if $u_{i-1} > v_{i-1}$ **then**

            $u_i \leftarrow u_{i-1}$

        else

            $u_i \leftarrow v_{i-1}$

    $R \leftarrow 0$

    for $i \leftarrow n$ **downto** 1 **do**

        $t \leftarrow u_i$

        $R \leftarrow R + d_{n-i,\,t} \cdot (v_i - 1)$

    $\text{Rank}(v) \leftarrow R$

end.

ALGORITHM 13: *Unrank Restricted Growth Function*

begin

    $u_1 \leftarrow 1$

    $v_1 \leftarrow 1$

    for $i \leftarrow 2$ **to** n **do**

        $t \leftarrow u_{i-1}$

        if $u_{i-1} \cdot d_{n-i,t} \le R$ **then**

            $v_i \leftarrow t + 1$

            $R \leftarrow R - t \cdot d_{n-i,t}$

            $u_i \leftarrow v_i$

        else

            $v_i \leftarrow \lfloor R/(d_{n-i,t}) \rfloor + 1$

            $R \leftarrow R \bmod d_{n-i,t}$

            $u_i \leftarrow t$

    $\text{Unrank}(R) \leftarrow v$

end.

Notes

The books by Nijenhuis and Wilf [N-W]; Reingold, Nievergelt and Deo [R-N-D]; and Williamson [Wi] contain more material in this area. Some of these books also explicitly give programs. It is possible to give loop-free versions of some of the listing algorithms in this chapter. For examples, see [Eh] or [Jo-Wh-Wi]. Gray codes are important in computer science and were first described in [Gra]. The version given here is called a binary reflected Gray code. Restricted growth functions were studied extensively by Milne [Mi]. Exercises 4 and 5 below are from [R-N-D].

Exercises

1. [1]   a) How many permutations lie between the permutations 1572634 and 7241365 in the Johnson-Trotter algorithm?

b) Find the millionth permutation (that is, Unrank(999999)) of [12] in the Johnson-Trotter algorithm.

2. [2C]   Give the listing algorithm for the lex list of permutations of [n] of length k. Upon what recurrence formula is this list based?

3. [3]   a) Find a Rank and Unrank formula for permutations listed in lex order.

b) Give a bijective proof of $1(1!) + 2(2!) + \ldots + n(n!) = (n+1)! - 1$.

c) Prove that any $0 \le m \le (n+1)! - 1$ can be uniquely expressed as

$$m = \sum_{i=1}^{n} a_i\, i!$$

where $0 \le a_i \le i$.

4. [3]   A permutation π is called *even* if π can be expressed as a product of an even number of transpositions. Suppose $\text{Rank}(\pi) = M$ in lex order, where $M = d_1 1! + d_2 2! + \ldots + d_{n-1}(n-1)!$, $0 \le d_j \le j$. Show that π is even if and only if $d_1 + d_2 + \ldots + d_{n-1}$ is even. What is the corresponding result for the Johnson-Trotter algorithm?

5. [3C]   Write a program to find, for various values of n, the number of permutations π of [n] which satisfy $\pi_i - i \equiv \pi_j - j \mod n$ implies $i = j$. This condition means that the numbers $\pi_j - j \mod n$, $1 \le j \le n$, are all distinct. An example of such a π is 321. Formulate and prove as many theorems as you can.

6. [2C]   Write a program to generate random permutations of n. Use it to estimate the answer to Exercise 5 for various values of n.

7. [2]   Show that for any given sequence $(a_1, \ldots, a_n)$, $0 \le a_i \le i - 1$, there is exactly one permutation whose inversion sequence is $(a_1, \ldots, a_n)$.

8. [2]   Give bijective proofs of

$$n \binom{n-1}{k-1} = k \binom{n}{k}$$

and

$$\binom{n+1}{k+1} = \sum_{m=k}^{n} \binom{m}{k}.$$

9. [1]   What is the rank of $(2, 3, 4, 7, 9)$ for $n = 9$ in the list generated by Algorithm 4? What is its successor?

10. [2]   Prove that if the k-element subsets of [n] are listed in colex order, and j is replaced by $n + 1 - j$ in each subset, the resulting list is lex order backwards. Use this to give a Rank and Unrank procedure for lex order. Do Exercise 9 for this list, where $n = 10$.

11. [2]   Show that $v_1 < v_2 < \ldots < v_k$ in Algorithm 6.

12. [3]   Let $A_0, A_1, \ldots, A_{m-1}$ be the first m k-element subsets of [n] in colex order. Let $\Pi = \{B : |B| = k - 1, B \subset A_i \text{ for some } 0 \le i \le m - 1\}$. If

$$m = \sum_{i=1}^{k} \binom{v_i - 1}{i}$$

and $v_i = i$ for $i = 1, 2, \ldots, r$, show that

$$|\Pi| = \sum_{i=r+1}^{k} \binom{v_i-1}{i-1}.$$

For example, if $n = 5$, $k = 3$ and $m = 5$, the first five subsets are 123, 124, 134, 234 and 125, $\Pi = \{12, 13, 23, 14, 24, 34, 15, 25\}$ and $|\Pi| = 8$.

13. [3C] Give the Rank and Unrank algorithms for Algorithm 7. Find Rank($3^3 2^3 1^4$).

14. [2]   Give a recursion for $p(n, k)$, the number of partitions of n whose largest part is $< k$. What is $p(n)$ in terms of $p(n, k)$?

15. [3C] Write a program to list all partitions of n with distinct parts. How many have an even (odd) number of parts? Formulate your conjectures and prove as many as you can.

16. [3C] Write a program which lists all partitions of n
        a) whose even parts are distinct;
        b) all of whose parts have multiplicity ≤ 3.
Formulate a conjecture and prove it.

17. [3C] Write a program which lists all partitions of n
        a) with an odd number of parts;
        b) with an even number of parts; and
        c) into distinct, odd parts.
Formulate a conjecture and prove it.

18. [3C] Write a program which lists all partitions of n
        a) into parts which are congruent to 1 or 4 mod 5;
        b) into parts whose differences are at least two; and
        c) into distinct parts, where each even part is $>$ twice the number of odd parts.
Investigate your data. If a) is replaced by 2 or 3 mod 5, can you find an appropriate b)?

19. [1]   What is the last n-tuple on the list $G(n)$?

20. [2]   Suppose $0 \leq i \leq 2^n - 1$, and let $i = a_{n-1}a_{n-2}...a_0$ be the base 2

representation of i. For the Gray code $G(n)$, what is $\text{Unrank}(i)$ in terms of $(a_{n-1}, a_{n-2}, \ldots, a_0)$?

21. [2]   Use Algorithms 9 and 10 to state and prove a representation theorem for integers i satisfying $0 \le i \le m_1 m_2 \cdots m_n - 1$.

22. [2]   Show that Algorithm 8 for $m_i = i$ produces the permutation list of Algorithm 1 by the inversion vector.

23. [2]   Give a bijective proof that the Bell numbers B_n satisfy

$$B_{n+1} = \sum_{k=0}^{n} \binom{n}{k} B_k.$$

24. [1]   Find $\text{Rank}(1231142)$ in the lex list of RG functions.

25. [2]   Give bijective proofs of

a) $$\binom{n}{k}\binom{k}{m} = \binom{n}{m}\binom{n-m}{k-m}$$

b) $$S(n+1, m+1) = \sum_{k=m}^{n} \binom{n}{k} S(k, m).$$

26. [2]   Give a bijective proof of Vandermonde's theorem

$$\sum_{k=0}^{i} \binom{m}{k}\binom{n}{i-k} = \binom{m+n}{i}.$$

27. [3]   What is the representation theorem for integers i satisfying $0 \le i \le B_n - 1$ which follows from Algorithms 12 and 13?

CHAPTER 2

Partially Ordered Sets

Partially ordered sets, or *posets*, appear in many branches of mathematics, but they are fundamental in combinatorics. For example, many of the important enumeration techniques (generating functions, inclusion-exclusion) have their theoretical foundation in some underlying poset.

In Chapter 1 we considered five different combinatorial objects. In this chapter we shall describe a poset for each of the five objects. For integer partitions we give two different posets, giving a total of six posets. The listing algorithms can be used to establish some non-trivial properties of these posets, as we see in §§2.2, 2.3 and 2.4. We concentrate on the Boolean algebra, and lex order from Chapter 1 will be crucial in developing its properties.

§2.1 Six Posets

A partially ordered set $(P, \leq)$ is a set P with an order relation $\leq$ which has the following properties:

(i) $a \leq a$ for all $a \in P$,

(ii) $a \leq b$ and $b \leq a$ implies $a = b$, and

(iii) $a \leq b$ and $b \leq c$ implies $a \leq c$.

We will consider only finite posets. A number of methods can be used to describe a poset: one is to maintain a list of all pairs (a, b) with $a < b$ (this means a $\leq b$ and $a \neq b$). We say that *b covers a* if $a < b$ and there is no c satisfying $a < c < b$. We shall write $a <\cdot b$ for b covers a. Because of property (iii), we need only maintain a list of all pairs (a, b) with $a <\cdot b$. These are called the *covering relations* of P.

We can visualize a poset as a graph with the "largest" elements of P as vertices at the top, the "smallest" at the bottom, and the other elements of P distributed appropriately in between. An edge connects a and b if and only if $a <\cdot b$.

As an example, suppose $P = \{a, b, c, d, e, f\}$ and the covering relations in P are

$$\{(b, a), (c, b), (e, a), (e, d), (f, c), (f, e)\}.$$

We can draw the poset as below.

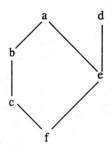

Such a diagram is called a *Hasse diagram*.

We now describe the posets for the objects of Chapter 1.

A. Permutations Although several orders are possible, we choose one which is closely related to Algorithm 1. Here are the covering relations. Suppose π and σ are permutations of $[n]$, and let $\pi = (\pi_1, \dots, \pi_n)$ in one-line notation. We say $\sigma <\cdot \pi$ if σ can be obtained by transposing π_i and π_{i+1}, where $\pi_i > \pi_{i+1}$. Thus, roughly speaking, π has more disorder than σ. As we move up this poset, we create more disorder. We call this poset the *inversion poset* and refer to it as $\mathscr{I}_n$. The Hasse diagram of $\mathscr{I}_4$ is given below.

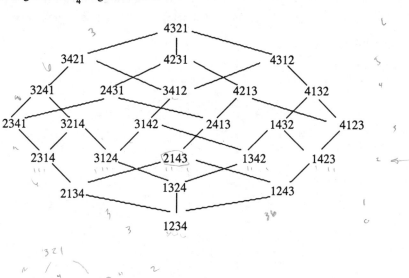

B. Subsets The subsets of [n] are naturally ordered by set inclusion. This poset is called the *Boolean algebra* (even though a Boolean algebra is a special kind of poset, of which this is one example). We denote it by B_n. The Hasse diagram for B_4 is given below.

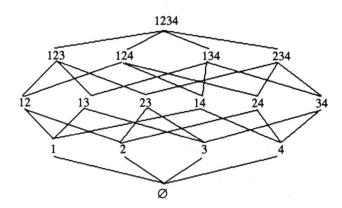

C. Set partitions The set partitions of [n] are ordered by reverse refinement. That is, $\pi = \{B_1, \dots, B_k\} <\cdot \{A_1, \dots, A_j\} = \sigma$ if $j = k + 1$ and for some p, m and t, $B_p = A_m \cup A_t$, with all of the other blocks of π and σ identical. Thus the covering relations are obtained by splitting one block into two blocks, for example $147\text{-}29\text{-}368\text{-}5 <\cdot 147\text{-}29\text{-}38\text{-}5\text{-}6$. (Note that we have dropped the set symbols { }.) Call this poset the *partition lattice* and denote it P_n. The Hasse diagram of P_4 is given below.

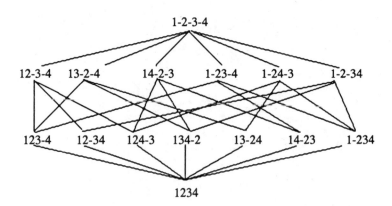

D. Integer Partitions   There are two common posets associated with integer partitions.

D1.   For the first one, take any partition λ and let P be the set of all partitions whose Ferrers diagram is contained in the Ferrers diagram of λ. Order these partitions by containment of their Ferrers diagram. This poset is called *Young's lattice* and is designated $\mathcal{Y}_\lambda$. The Hasse diagram of $\mathcal{Y}_{332}$ is given below.

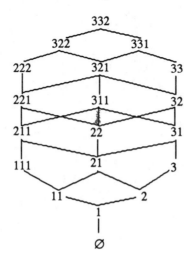

D2.   The other order is called *dominance*. Let P be all integer partitions of n. We say that $\lambda = (\lambda_1, \dots , \lambda_k)$ *dominates* $\mu = (\mu_1, \dots , \mu_j)$ if $\lambda_1 + \dots + \lambda_t \geq \mu_1 + \dots + \mu_t$ for all $t \geq 1$. (We append 0 parts to λ and μ for $t > k$ or $t > j$.) The poset is called the *domination lattice* and is denoted $\mathcal{D}_n$. The Hasse diagram of $\mathcal{D}_7$ is given below.

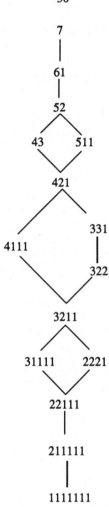

E. Product Spaces Since there is a bijection between all subsets of $[n]$ and the product space $\{0, 1\}^n$, the Boolean algebra B_n can be identified with $\{0, 1\}^n$. We generalize this to $P = \{0, 1, \dots, m-1\}^n$. Given two n-tuples in P, $v = (v_1, \dots, v_n)$ and $w = (w_1, \dots, w_n)$, we say that $v <\cdot w$ if v and w agree in all but one entry, and in that entry $v_i + 1 = w_i$. We call this poset a *product of chains*, and denote it C_{nm}. The Hasse diagram for C_{23} is given below.

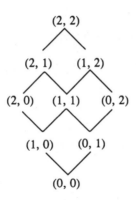

$(2, 2)$

$(2, 1)$ $(1, 2)$

$(2, 0)$ $(1, 1)$ $(0, 2)$

$(1, 0)$ $(0, 1)$

$(0, 0)$

We now give several important properties of a general poset P. We may then ask which of our six posets has each of these properties.

(1) Ranked A poset P is called *ranked* if each element $a \in P$ can be assigned a non-negative integer, rank(a), so that $a <\cdot b$ implies rank(b) = rank(a) + 1. The set $L_i = \{a \in P : \text{rank}(a) = i\}$ is called the set of rank i, or a *level set*. This rank function should not be confused with the Rank function of Chapter 1. The inversion poset $\mathcal{J}_n$ is ranked by the number of *inversions* of π, $|\{(i, j) : i < j \text{ and } \pi_i > \pi_j\}|$. Clearly this number is the sum of the entries of the inversion vector of π. In the other posets, $\mathcal{B}_n$ is ranked by the size of the subset, $\mathcal{P}_n$ by the number of blocks, $\mathcal{Y}_\lambda$ by the integer being partitioned and $\mathcal{C}_{nm}$ by the sum of the entries in the n-tuple. The poset $\mathcal{D}_n$ is not ranked for $n \geq 7$.

(2) Rank unimodal A finite ranked poset P is called *rank unimodal* if $w_i = |L_i|$, $w_i = 0$ for $i > m$ and $w_0 \leq w_1 \leq \ldots \leq w_k \geq w_{k+1} \geq \ldots \geq w_m$ for some $k \leq m$. It is clear that the Boolean algebra $\mathcal{B}_n$ is rank unimodal. It can be shown that $\mathcal{J}_n$, $\mathcal{P}_n$ and $\mathcal{C}_{nm}$ are also rank unimodal. Unimodality is often difficult to prove. The case of $\mathcal{Y}_\lambda$ for arbitrary λ is unsettled although for $\lambda = n^m$, Young's lattice is rank unimodal.

(3) Rank symmetric A ranked poset P is called *rank symmetric* if there is an m such that $w_i = 0$ for $i > m$ and $w_i = w_{m-i}$ for $0 \leq i \leq m$. The posets $\mathcal{J}_n$, $\mathcal{C}_{nm}$ and $\mathcal{B}_n$ are rank symmetric, while $\mathcal{P}_n$, $\mathcal{Y}_\lambda$ and $\mathcal{D}_n$ are not in general.

(4) Order symmetric A poset P is called *order symmetric* if the poset $\tilde{P}$ obtained by reversing the order of P is isomorphic to P. Another way of saying this is that if the Hasse diagram of P is turned upside down, it looks the same. The posets $\mathcal{A}_n$, $\mathcal{B}_n$, $\mathcal{C}_{nm}$ and $\mathcal{D}_n$ are order symmetric, while $\mathcal{P}_n$ and $\mathcal{Y}_\lambda$ are not.

(5) Lattice A poset P is called a *lattice* if every pair $\{a, b\}$ of elements of P has a least upper bound (called the *join* of a and b, $a \vee b$) and a greatest lower bound (called the *meet* of a and b, $a \wedge b$). A *least upper bound* (*greatest lower bound*) of $\{a, b\}$ is an element $c \in P$ which is an *upper* (*lower*) *bound* of $\{a, b\}$, $a \leq c$ and $b \leq c$, and below (above) all other upper (lower) bounds. The following poset is <u>not</u> a lattice.

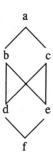

Each of the six posets is a lattice.

(6) Existence of maximum and minimum elements We say a poset $(P, \leq)$ has a *minimum* element $\hat{0}$ if $\hat{0} \in P$ and $\hat{0} \leq a$ for all $a \in P$. We say P has a *maximum* element $\hat{1}$ if $\hat{1} \in P$ and $a \leq \hat{1}$ for all $a \in P$. All of our six posets have maximum and minimum elements.

(7) Sperner property If $(P, \leq)$ is a poset, a subset $S \subset P$ is called *independent* if for all a and $b \in S$, neither $a \leq b$ nor $b \leq a$. Sometimes such a set is called an *antichain*. For a ranked poset, it is clear that the level sets L_i are independent sets. A finite ranked poset P has the *Sperner property* if the size of the largest independent set is $\max\{w_i : i \geq 0\}$. This means that we can do no better than to take the largest level set as our independent set.

For the Boolean algebra, the largest level number is the middle binomial coefficient. A theorem called Sperner's Theorem states that the size of the largest independent set of $\mathcal{B}_n$ is this binomial coefficient. So $\mathcal{B}_n$ has the Sperner property. We shall prove this in the next section using lex order. It can also be shown that $\mathcal{C}_{nm}$ has the Sperner property.

In the other posets the situation is less clear. It was conjectured by Rota in 1968 that $\mathcal{P}_n$ had the Sperner property, but 10 years later Canfield [Ca], using asymptotic methods, showed that it does not. More recent research has shown that the first n for which P_n does not have the Sperner property is $\leq N$, where $N \approx 1000$.

The case of $\mathcal{L}_n$ is unsettled. For a general λ, the case of $\mathcal{Y}_\lambda$ is also open. However, for $\lambda = n^m$ (the Ferrers diagram of λ is an $n \times m$ rectangle), $\mathcal{Y}_\lambda$ has the Sperner property. This is a recent theorem of Stanley ([Sta1]), although it is implicit in Pouzet [Po].

We summarize these results in the table below. You are asked to verify several of these entries in the exercises. (A complete list of the pertinent exercises is given in the Notes.)

	Ranked	Rank Unimodal	Rank Symmetric	Order Symmetric	Lattice	$\hat{0}$	$\hat{1}$	Sperner Property
$\mathcal{L}_n$	yes	yes	yes	yes	yes	yes	yes	?
B_n	yes	yes	yes	yes	yes	yes	yes	yes
P_n	yes	yes	no	no	yes	yes	yes	no
$\mathcal{Y}_\lambda$	yes	?	no	no	yes	yes	yes	?
$\mathcal{D}_n$	no	no	no	yes	yes	yes	yes	no
C_{nm}	yes	yes	yes	yes	yes	yes	yes	yes

§2.2 Matching in the Boolean algebra

In this section we shall prove that the Boolean algebra B_n has the Sperner property. We shall use the lex order list of subsets given in Chapter 1. The basic idea of the proof is that lex order naturally gives a matching in B_n. This matching gives a decomposition of B_n into chains. From the chain decomposition of B_n it is easy to establish the Sperner property. First we need to define these terms.

A *chain* C in a finite poset $(P, \leq)$ is a sequence of elements $a_1 < a_2 < \ldots < a_m$. The *length* of the chain is m. For example, $\emptyset \subset \{1, 3\} \subset \{1, 2, 3, 5\}$ is a chain in B_5. Two chains are called *disjoint* if they have no elements in common. A *chain decomposition* of P is a partitioning of P into a set of disjoint chains, $\{C_i\}$. Clearly, we could let each C_i have exactly one element and have a chain decomposition of P into $|P|$ chains. The next theorem says that the existence of special chain decompositions of P implies the Sperner property.

THEOREM 2.1 *Let* P *be a finite ranked poset with a largest level set* L_k. *If* P *has a chain decomposition* $\{C_i\}$, *and each* C_i *contains one element of* L_k, *then* P *has the Sperner property.*

Proof Let M be the size of a largest independent set S. Every element of S must be on a different chain of $\{C_i\}$. Thus, $M \le |\{C_i\}|$. Since each C_i has exactly one element of L_k, $|\{C_i\}| = |L_k|$.

One way to produce such a chain decomposition is to construct a match between adjacent level sets. Suppose $|L_i| \le |L_{i+1}|$. A function $f : L_i \to L_{i+1}$ is called a *matching* from L_i to L_{i+1} if f is an injection (this means $f(a) = f(b)$ implies $a = b$) and $a \le f(a)$ for all $a \in L_i$. Similarly define a matching from L_{i+1} to L_i if $|L_{i+1}| \le |L_i|$.

THEOREM 2.2 *Suppose* P *is rank unimodal and there is a matching between any two adjacent levels of* P. *Then* P *has the Sperner property.*

Proof Since P is rank unimodal, $|L_1| \le |L_2| \le \ldots \le |L_k| \ge \ldots \ge |L_m|$ for some k. Let $f_i : L_{i-1} \to L_i$ be a matching for $i = 1, 2, \ldots, k$, and let $g_i : L_i \to L_{i-1}$ be a matching for $i = k + 1, \ldots, m$. Delete from the Hasse diagram of P all edges except those of the form $\{a, f_{i+1}(a)\}$ or $\{a, g_i(a)\}$. The new Hasse diagram is a union of chains, each of which must contain an element of L_k. These chains are naturally also chains in P, so Theorem 2.1 implies that P has the Sperner property.

The proof of Theorem 2.2 gives us an algorithm to construct the chain decomposition of P which satisfies the hypothesis of Theorem 2.1. We merely identify those chains in the proof. Suppose $L_0 \ne \varnothing$, and choose any $a \in L_0$. Let $a_0 = a$, $a_1 = f_1(a_0)$, $\ldots$, $a_k = f_k(a_{k-1})$. We now have a chain $a_0 <\cdot a_1 <\cdot \ldots <\cdot a_k$. If a_k is in the image of g_{k+1}, let a_{k+1} be the unique element in L_{k+1} such that $g_{k+1}(a_{k+1}) = a_k$. We continue in this manner until either

(a)   we encounter an a_i which is not in the image of g_{i+1}, or

(b)   we reach $a_m \in L_m$.

Let $C_1 = a_0 <\cdot a_1 <\cdot \ldots <\cdot a_t$ be this chain. Now repeat the preceding construction until the elements of L_0 are exhausted. Then continue the construction by choosing $a \in L_1 - f_1(L_0)$ and moving up the chain of matchings. This process is continued at all levels. After we complete this process at level k, we have obtained the appropriate chain decomposition.

Next we apply Theorem 2.2 to the Boolean algebra $\mathcal{B}_n$. We need a matching between any two levels, L_p to L_{p+1}. Let $A_1, A_2, \ldots$ be the p-element subsets of [n] listed in lex order. Let $B_1, B_2, \ldots$ be the (p+1)-element subsets of [n] listed in lex order. How can we match these sets? One idea would be to match A_1 to the first B_j such that $A_1 \subset B_j$. Then match A_2 to the first unmatched B_j such that $A_2 \subset B_j$, and continue. This is our next algorithm.

The set U is the collection of (p+1)-element subsets which have been matched. The subroutines GetFirstSubset(k, A, b) and GetNextSubset(k, A, b) return the first and next k-element subset in lex order. The boolean variable b is returned true if the list is complete and false otherwise. This algorithm will list all the matched pairs and all unmatched p- and (p+1)-element subsets.

ALGORITHM 14: *Match to First Available*

begin
    $U \leftarrow \varnothing$
    GetFirstSubset(p, A, EndOfList)
    while not EndOfList **do**
        GetFirstSubset(p + 1, B, ListDone)
        StopLoop $\leftarrow$ ListDone
        while not StopLoop **do**
            if $A \subset B$ **and** $B \notin U$ **then**
                StopLoop $\leftarrow$ **true**
            else
                GetNextSubset(p + 1, B, ListDone)
                StopLoop $\leftarrow$ ListDone
        if not ListDone **then**
            PrintMatch(A, B)
            $U \leftarrow U \cup \{B\}$
        else
            PrintNoMatch(A)
        GetNextSubset(p, A, EndOfList)
    GetFirstSubset(p + 1, B, EndOfList)
    while not EndOfList **do**
        if $B \notin U$ **then**

PrintNoMatch(B)

GetNextSubset(p + 1, B, EndOfList)

end.

It would seem remarkable that Algorithm 14 would work, that is, the function f would be a matching. Yet the next theorem says that it does.

THEOREM 2.3 *If* $p < \lceil n/2 \rceil$, *then f is a matching from* L_p *to* L_{p+1} *in* B_n. *If* $p \geq \lceil n/2 \rceil$, *then* f^{-1} *is a matching from* L_{p+1} *to* L_p *in* B_n.

Proof We approach this theorem by defining a completely new function ϕ which satisfies its conclusion. Then we show that $\phi = f$.

We shall write our p-element subsets of [n] as n-tuples of 0's and 1's as described in Chapter 1. These n-tuples can be pictured as lattice paths from $(0, 0)$ to $(n, 2p - n)$. Each digit in the n-tuple represents a step one unit to the right, and one unit either up (a "1") or down (a "0"). For example, if

$$A = \{1, 3, 4, 6, 7, 10, 12, 13, 16, 19\} \subset [21],$$

we write it as (101101100101100100100) or graph it as the following.

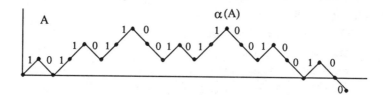

Clearly there is a bijection between all such lattice paths from $(0, 0)$ to $(n, 2p - n)$ and the p-element subsets of [n]. Henceforth, we shall refer to this lattice path (or graph) as the subset A itself.

Let $\alpha(A) = (\alpha_x(A), \alpha_y(A))$ be the rightmost peak (absolute maximum) in the graph of A. In our example above, $\alpha(A) = (13, 3)$. If $\alpha_x(A) \neq n$, then the edge immediately to the right of $\alpha(A)$ must correspond to a "0". Let $\phi(A)$ be the (p+1)-element subset of [n] obtained by changing that "0" to a "1". Note that to the left of $\alpha(A)$, the graphs of A and $\phi(A)$ coincide. To the right of $\alpha(A)$, the graphs have the same relative position, but the graph of $\phi(A)$ has been "lifted" 2 units.

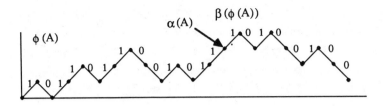

We now make a number of observations about ϕ. We assume $n \geq 1$.

(1)   If $p < \lceil n/2 \rceil$, then ϕ is well-defined. For if n is even, $p < n/2$ implies $2p-n < 0$ and the graph of A ends below the x-axis, and $\alpha_x(A) \neq n$. If n is odd, $p < (n+1)/2$ implies $2p-n < 1$. Since n is odd, $2p-n \neq 0$, thus $2p-n < 0$ and the argument for n even applies.

(2)   If $p \geq \lceil n/2 \rceil$, then $\phi(A)$ is not defined if and only if $\alpha_x(A) = n$.

(3)   For a subset B, let $\beta(B) = (\beta_x(B), \beta_y(B))$ be the leftmost peak of B. Suppose that $B = \phi(A)$ for some A. Then $\alpha(A)$ and $\beta(B) = \beta(\phi(A))$ are the left and right hand endpoints of the same edge. This is because $\alpha_y(A) \geq$ y-coordinate of every point left of $\alpha(A)$, and $\alpha_y(A) >$ y-coordinate of every point right of $\alpha(A)$. Then in $\phi(A)$, the y-coordinate of the vertex to the right of $\alpha(A)$ must be $>$ y-coordinate of every point to its left and $\geq$ y-coordinate of every vertex to its right. The reader is encouraged to look at some examples and verify these remarks.

(4)   It follows then that if $p \geq \lfloor n/2 \rfloor$ and $|B| = p + 1$, $\phi^{-1}(B)$ will exist and can be found by converting the edge immediately to the left of $\beta(B)$ from a "1" to a "0". The only case where this cannot be done is when $\beta_x(B) = 0$. But if n is even, $p \geq n/2$ implies $2p-n \geq 0$, so $2(p+1)-n \geq 2$. If n is odd, $p \geq (n-1)/2$ implies $2(p+1)-n \geq 1$. In either case, $(0,0) \neq \beta(B)$.

(5)   If $p < \lfloor n/2 \rfloor$ and $|B| = p + 1$, then $\phi^{-1}(B)$ does not exist if and only if $\beta(B) = (0,0)$.

These observations prove that if $p < \lceil n/2 \rceil$, ϕ is a matching from L_p to L_{p+1}; and if $p \geq \lceil n/2 \rceil$, ϕ^{-1} is a matching from L_{p+1} to L_p. To complete the proof of Theorem 2.3, we must show that $\phi = f$.

We proceed by induction in the $p < \lceil n/2 \rceil$ case. The $p \geq \lceil n/2 \rceil$ case can be done similarly.

The first A in lex order can easily be shown to satisfy $f(A) = \phi(A)$. So we can assume by induction that $\phi(\tilde{A}) = f(\tilde{A})$, for all $\tilde{A}$ preceding A in lex order. We know that there is at least one B available as a match for A, namely $\phi(A)$. We need not consider any B's further down the lex order list than $\phi(A)$, for Algorithm 14 would select $\phi(A)$ before them. So we assume that there is a B which precedes $\phi(A)$ in lex order, which contains A, and which is still free ($f^{-1}(B)$ does not yet exist).

Given such a B, let γ denote the right endpoint of the edge e changed from a "0" to a "1" to convert A to B. Since $B \neq \phi(A)$, $\gamma \neq \beta(B)$, and we can consider two cases.

Case 1: γ precedes $\beta(B)$.

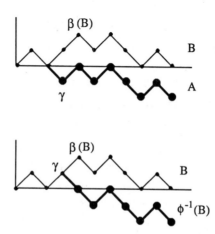

Then A and $\phi^{-1}(B)$ will agree up to the edge before γ, but A will have a "0" along this edge while $\phi^{-1}(B)$ will have a "1". So in lex order, $\phi^{-1}(B)$ will precede A. By induction, $\phi^{-1}(B)$ has already been matched to $f(\phi^{-1}(B)) = \phi(\phi^{-1}(B))$ $= B$, which implies that B is not free, a contradiction.

Case 2: γ follows $\beta(B)$. The graphs of B and $\phi(A)$ differ at two places: to the left of γ, ($\phi(A)$ has a "0", B has a "1"), and to the right of $\alpha(A)$, ($\phi(A)$ has a "1", B has a "0"). First we show that $\alpha(A)$ must precede γ. Suppose not, so $\alpha(A)$ follows γ. Then B and $\phi(A)$ agree up to the edge e before γ. The sets A and $\phi(A)$ differ only at the edge following $\alpha(A)$. (See the diagram below.)

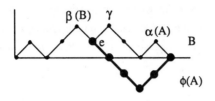

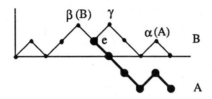

Since B has a "1" at e while A has a "0", the graph of B must lie strictly above the graph of A between γ and $\alpha(A)$. The peak $\beta(B)$ of B must therefore lie above the graph of A at $\alpha(A)$. But γ follows $\beta(B)$ so that B agrees with A at $\beta(B)$. This shows that $\alpha(A)$ is not a peak of A, a contradiction.

Since we have just shown that $\alpha(A)$ precedes γ, $\phi(A)$ precedes B in lex order. This contradicts our hypothesis that B precedes $\phi(A)$ in lex order. This completes the proof that $\phi = f$, and thus also completes the proof of Theorem 2.3.

§2.3 The Littlewood-Offord Problem

Littlewood and Offord [Li-O] asked the following question. Let $p_n(x)$ be a polynomial of degree n with complex coefficients,

$$p_n(x) = \sum_{k=0}^{n} a_k x^k.$$

Certainly $p_n(x)$ has n complex roots, and a smaller number of real roots. Now consider the 2^n polynomials obtained from $p_n(x)$ by arbitrarily changing the signs of a_k, $1 \leq k \leq n$. What is the average number of real roots of these 2^n polynomials? Littlewood and Offord showed that this average was rather small. In this section we discuss a combinatorial problem which arose naturally in this context. The complete solution to this problem is due to Kleitman (see [Gre-K1]). It uses the chain decomposition of the Boolean algebra given in §2.2.

Before stating the problem, we need another property of the chain decomposition. Let P be a rank unimodal and rank symmetric poset, with ranks from 0 to m. We call a chain decomposition $\{C_i\}$ of P a *symmetric chain decomposition* if each chain C_i has the form

$$C_i : a_j <\cdot a_{j+1} <\cdot \ldots <\cdot a_{m-j},$$

for some j, where $\text{rank}(a_i) = i$ for all i, $j \le i \le m - j$. This says that the chains are symmetric with respect to the middle of the poset. The matching in B_n produces the chain decomposition $\{C_1, C_2, C_3\}$,

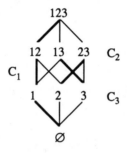

where

$$C_1 : \varnothing \subset \{1\} \subset \{1, 2\} \subset \{1, 2, 3\},$$
$$C_2 : \{2\} \subset \{2, 3\},$$
and $$C_3 : \{3\} \subset \{1, 3\}.$$

The matching f always gives a symmetric chain decomposition of B_n. You are asked to prove this in Exercise 25.

The Littlewood-Offord problem is the following. Let $\{v_1, \ldots, v_n\}$ be a multiset (repetitions are allowed) of vectors from R^N such that $\|v_i\| \ge 1$ for all i. Given a subset $A \subset [n]$, define a new vector

$$w_A = \sum_{i \in A} v_i .$$

Clearly, as a multiset, there are 2^n such vectors w_A. How many such vectors can lie in any sphere of diameter 1? The answer is the following theorem.

THEOREM 3.1 *Let S be any sphere of diameter* 1. *Then the number of vectors* w_A

from the origin which terminate inside S *is* $\leq M(n)$, *where*

$$M(n) = \binom{n}{\lfloor n/2 \rfloor}.$$

Moreover, this bound is best possible.

Proof First we show that, in general, the bound cannot be decreased. Suppose all of the vectors are the same, $v_i = v$. Then $w_A = |A|\, v$, and if $|A| = \lfloor n/2 \rfloor$ is fixed, w_A is always the same vector. So the sphere S centered at $\lfloor n/2 \rfloor v$ contains $M(n)$ vectors. It does not contain any of the other w_A because $\|v\| \geq 1$.

The proof will mimic an inductive construction of the symmetric chain decomposition of B_n, first given by deBruijn, Tengbergen and Kruyswijk [deB-T-K]. Suppose $\{C_i\}$ is a symmetric chain decomposition of B_n. We need to modify the subsets in these chains to create a symmetric chain decomposition of B_{n+1}. Let $\tilde{C}_i$ be the chain in B_{n+1} which results from adding $n+1$ to each subset of C_i. Unfortunately, $\tilde{C}_i$ is not a symmetric chain in B_{n+1}. Let α_i be the top subset of $\tilde{C}_i$, and put $D_i = \tilde{C}_i - \alpha_i$ and $E_i = C_i \cup \alpha_i$. It is easy to see that D_i and E_i are both symmetric chains in B_{n+1}. We claim that $\{D_i\} \cup \{E_i\}$ is a symmetric chain decomposition of B_{n+1}.

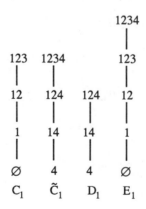

Let β be any subset of $[n+1]$. If $n+1$ is not in β, then $\beta \in C_i$ for some i, so $\beta \in E_i = C_i \cup \alpha_i$ also. Otherwise $n+1 \in \beta$, and $\beta - \{n+1\} \in C_i$ for some i. If $\beta - \{n+1\}$ is the top subset of C_i, then $\beta = \alpha_i$ and $\beta \in E_i = C_i \cup \alpha_i$. If $\beta - \{n+1\}$ is not the top subset of C_i, then $\beta \in \tilde{C}_i - \alpha_i = D_i$. So we have proven that all subsets β of $[n+1]$ lie in one of the chains in $\{D_i\} \cup \{E_i\}$. Clearly these chains are disjoint.

We now return to the Littlewood-Offord problem. The idea is to decompose the 2^n vectors w_A into $M(n)$ blocks $\{B_i\}$, $1 \leq i \leq M(n)$, such that the distance between any two vectors within a block B_i is ≥ 1. Then any sphere S of diameter 1 could contain, at most, one vector from each B_i.

The blocks $\{B_i\}$ are built inductively just as the chains $\{C_i\}$ were built. Suppose blocks $\{B_i\}$, $1 \leq i \leq M(n)$, have been chosen for $\{v_1, \ldots, v_n\}$. We define new blocks for for the vectors $\{v_1, \ldots, v_{n+1}\}$. The vector v_{n+1} takes the place of $n+1$ in the symmetric chain decomposition of $\mathcal{B}_n$. Let $B_i = \{t_1, \ldots, t_p\}$. We let $\tilde{B}_i = \{t_1 + v_{n+1}, \ldots, t_p + v_{n+1}\}$, which is analogous to the definition of $\tilde{C}_i$. Next we need a vector analogue of α_i, the top subset of $\tilde{C}_i$. This is the vector $t_m + v_{n+1}$, where t_m is a vector of B_i with maximal component in the v_{n+1} direction. Our two new blocks are $D_i = \tilde{B}_i - \{t_m + v_{n+1}\}$ and $E_i = B_i \cup \{t_m + v_{n+1}\}$. It is clear that $\{D_i\}$ and $\{E_i\}$ partition the set of vectors w_A, $A \subset [n+1]$. The total number of blocks is exactly the number of chains in the symmetric chain decomposition of $\mathcal{B}_{n+1}$, $M(n+1)$. It remains to prove the distance condition for any two vectors from a given block.

By induction, we need only check the vectors $t_m + v_{n+1}$ and t_i of E_i. Let

$$(3.1) \qquad t_i = c_i\, v_{n+1} + w_i, \quad 1 \leq i \leq p,$$

where c_i is real and $w_i \perp v_{n+1}$. By our choice of t_m, $c_m \geq c_i$ for all i. So

$$(3.2) \qquad \|t_m + v_{n+1} - t_i\|^2 \;=\; \|(c_m - c_i + 1)\, v_{n+1} + w_m - w_i\|^2$$

$$= (c_m - c_i + 1)^2\, \|v_{n+1}\|^2 + \|w_m - w_i\|^2$$

$$\geq \|v_{n+1}\|^2 + \|w_m - w_i\|^2$$

$$\geq \|v_{n+1}\|^2 \geq 1.$$

To complete the proof, we must check the $n = 1$ case. Clearly the block $\{0, v_1\}$ works because $\|v_1\| \geq 1$.

Theorem 3.1 does not depend upon the dimension of the vector space R^N in which the vectors lie. It even holds for infinite dimensional vector spaces.

§2.4 Extremal Set Theory

Colex order of the k-element subsets has certain desirable properties which allow us to prove some important theorems in extremal set theory. In this section we shall prove two of these theorems: the Kruskal-Katona Theorem and the Erdös-Ko-Rado Theorem. Most of this section is devoted to proving the Kruskal-Katona Theorem. The Erdös-Ko-Rado Theorem is a corollary of the Kruskal-Katona Theorem.

First we begin with some notation. Let $\mathcal{F}$ be a collection of k-element subsets. Let $\partial\mathcal{F}$ denote the collection of all (k−1)-element subsets which are subsets of members of $\mathcal{F}$, that is, all of the (k−1)-element subsets which are covered by members of $\mathcal{F}$ in $\mathcal{B}_n$, for an appropriate n. We ask: how small can $|\partial\mathcal{F}|$ be for a given $|\mathcal{F}|$? As an example, take k = 4 and let $\mathcal{F}$ be these subsets.

$$1357$$
$$2357$$
$$2457$$
$$3458$$
$$2358$$

Then $\partial\mathcal{F}$ contains the following subsets.

135	257	358
137	245	458
157	247	238
357	457	258
235	345	
237	348	

You were asked in Exercise 12 of Chapter 1 to find $\partial\mathcal{F}$ for a special collection $\mathcal{F}$. We state that result slightly differently here as a lemma.

LEMMA 4.1 *Suppose that* $\mathcal{F}$ *consists of the first* m *k-element subsets listed in colex order, and that* $\text{Rank}(a_1, \dots, a_k) = m$ *so that*

$$m = \binom{a_k - 1}{k} + \dots + \binom{a_i - 1}{i}$$

with $a_j = j$ *for* $1 \le j \le i - 1$. *Then* $\partial \tilde{\mathcal{F}}$ *consists of the first* $\tilde{m}$ *(k–1)-element subsets listed in colex order, where*

$$\tilde{m} = \binom{a_k - 1}{k - 1} + \cdots + \binom{a_i - 1}{i - 1}.$$

Proof For example, if $k = 4$ and $m = 6$ the first 6 such subsets are

1234
1235
1245
1345
2345
1236.

The next is 1246 for which Rank(1246) = 6. From Theorem 2.1 of Chapter 1 we know that

$$m = \binom{5}{4} + \binom{3}{3}$$

so Lemma 4.1 implies that

$$\tilde{m} = \binom{5}{3} + \binom{3}{2} = 13.$$

Here is $\partial \tilde{\mathcal{F}}$:

123	135	345
124	235	126
134	145	136
234	245	236
125		

Note that n need not be given to list the first m k-element subsets in colex order.

For the proof, note that $\mathcal{F}$ consists of all of the k-element subsets of $[a_k - 1]$, all of the (k–1)-element subsets of $[a_{k-1} - 1]$ with a_k adjoined, all of the (k–2)-element subsets of $[a_{k-2} - 1]$ with a_k and a_{k-1} adjoined, etc. Thus, $\partial \mathcal{F}$ must consist of all of the (k–1)-element subsets of $[a_k - 1]$, all of the (k–2)-element subsets of $[a_{k-1} - 1]$ with a_k adjoined, all of the (k–3)-element subsets of

$[a_{k-2} - 1]$ with a_k and a_{k-1} adjoined, etc. The size of $\partial\mathcal{F}$ is clearly $\tilde{m}$, as claimed.

Because of Lemma 4.1, it is clear that Lemma 4.2 is equivalent to Theorem 4.3, which is the Kruskal-Katona Theorem.

LEMMA 4.2 *Suppose $\mathcal{F}$ is a collection of k-element subsets, and $\mathcal{A}$ is the collection of the first $|\mathcal{F}|$ k-element subsets in colex order. Then $|\partial\mathcal{A}| \leq |\partial\mathcal{F}|$.*

THEOREM 4.3 *If $\mathcal{F}$ is a collection of k-element subsets, where*

$$|\mathcal{F}| = \binom{a_k - 1}{k} + \dots + \binom{a_i - 1}{i}$$

then

$$|\partial\mathcal{F}| \geq \binom{a_k - 1}{k-1} + \dots + \binom{a_i - 1}{i-1}.$$

We prove Theorem 4.3 at the end of the section. Next we state the Erdös-Ko-Rado Theorem.

THEOREM 4.4 *Suppose that $k \leq \lfloor n/2 \rfloor$. Then the size of the largest collection $\mathcal{F}$ of k-element subsets of [n], no pair of which are disjoint, is*

$$\binom{n-1}{k-1}.$$

For example, if $n = 6$ and $k = 3$, we can choose our collection $\mathcal{F}$ to be

123	234	235	345
124	125	145	
134	135	245.	

But if we add any other 3-element subset to $\mathcal{F}$, it will be disjoint with one of the members of $\mathcal{F}$.

Proof of the Erdös-Ko-Rado Theorem Let $\mathcal{F}$ be such a collection of subsets, and suppose that

(4.1) $|\mathcal{F}| > \dbinom{n-1}{k-1}.$

Let $\overline{\mathcal{F}}$ be the collection of complements of the subsets of $\mathcal{F}$. Certainly $\overline{\mathcal{F}}$ consists of $|\mathcal{F}|$ $(n-k)$-element subsets of $[n]$. Furthermore, no member of $\mathcal{F}$ can be a subset of any member of $\overline{\mathcal{F}}$. (Suppose $A \in \mathcal{F}$, $B \in \overline{\mathcal{F}}$ and $A \subset B$. Then $A \cap \overline{B} = \emptyset$, and A and $\overline{B}$ are two members of $\mathcal{F}$ which would be disjoint.) The picture below describes this situation, since $k \le n - k$.

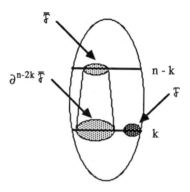

We now obtain a lower bound on the number of k-element subsets which lie below $\overline{\mathcal{F}}$. If we apply the map ∂ $n-2k$ times to $\overline{\mathcal{F}}$ (call this iterated map ∂^{n-2k}), we obtain all of these subsets. Let

(4.2) $|\overline{\mathcal{F}}| = \dbinom{a_{n-k}-1}{n-k} + \dots + \dbinom{a_i-1}{i}$

Since

(4.3) $|\overline{\mathcal{F}}| = |\mathcal{F}| > \dbinom{n-1}{k-1} = \dbinom{n-1}{n-k},$

a_{n-k} must be n. So by the Kruskal-Katona Theorem (Theorem 4.3),

(4.4) $|\partial \overline{\mathcal{F}}| > \dbinom{n-1}{n-k-1}.$

Repeating this $n - 2k - 1$ more times gives

(4.5) $\quad |\partial^{n-2k}\overline{\mathcal{F}}| \quad > \quad \left(\begin{array}{c} n-1 \\ n-k-(n-2k) \end{array}\right) = \left(\begin{array}{c} n-1 \\ k \end{array}\right).$

But $\mathcal{F}$ and $\partial^{n-2k}\overline{\mathcal{F}}$ have no elements in common, so

(4.6) $\quad |\mathcal{F}| \;+\; |\partial^{n-2k}\overline{\mathcal{F}}| \;\leq\; \left(\begin{array}{c} n \\ k \end{array}\right).$

Clearly (4.6), (4.5) and (4.1) contradict Pascal's triangle (§1.2, Eq. (2.3)) for the binomial coefficients.

Proof of the Kruskal-Katona Theorem The proof will be by induction on n, the size of the base set $U = \{x \in A : A \in \mathcal{F}\}$, and will have three steps.

(1)   Change $\mathcal{F}$ into a collection $\mathcal{F}'$ such that $|\partial\mathcal{F}'| \leq |\partial\mathcal{F}|$ and $|\mathcal{F}'| = |\mathcal{F}|$. The subsets in $\mathcal{F}'$ will contain 1 as frequently as possible.

(2)   Decompose the collection $\mathcal{F}'$ into two subcollections: those subsets with 1, $\mathcal{F}'(1)$, and those subsets without 1, $\mathcal{F}'(\tilde{1})$, so that $\mathcal{F}' = \mathcal{F}'(1) \cup \mathcal{F}'(\tilde{1})$. Derive a lower bound for $|\partial\mathcal{F}'|$ in terms of $\mathcal{F}'(1)$.

(3)   Use the induction hypothesis by deleting 1 from the subsets in $\mathcal{F}'(1)$ to simplify the lower bound of $|\partial\mathcal{F}'|$ from (2). From (1), this number is also a lower bound of $|\partial\mathcal{F}|$, and it turns out to be the bound of Theorem 4.3.

Step (1)   Given an integer $j, j \geq 2$, and a collection of subsets $\mathcal{A}$, we define a switching map S_j on $\mathcal{A}$. For $A \in \mathcal{A}$, let $S_j(A) = (A - \{j\}) \cup \{1\}$ if $j \in A$, $1 \notin A$ and $(A - \{j\}) \cup \{1\} \notin \mathcal{A}$. Otherwise we put $S_j(A) = A$. Clearly the new collection $S_j(\mathcal{A})$ has at least as many sets with 1 as $\mathcal{A}$ did. The next lemma tells us how S_j interacts with ∂.

LEMMA 4.5   *For any* $j, j \geq 2$, $\partial(S_j(\mathcal{F})) \subset S_j(\partial\mathcal{F})$.

Proof We must show that $\partial(S_j(A)) \subset S_j(\partial\mathcal{F})$ for all $A \in \mathcal{F}$.

   If $S_j(A) = A$, this is $\partial A \subset S_j(\partial\mathcal{F})$. If $j \notin A$, j is not in any of the sets in ∂A so $\partial A = S_j(\partial A) \subset S_j(\partial\mathcal{F})$. If 1 is in A, all but one $(A - \{1\})$ of the sets in ∂A contain 1, and the previous argument shows each of them is in $S_j(\partial\mathcal{F})$. For $A - \{1\}$, note that we can assume that j is in A, so that ∂A contains both $A - \{1\}$

and $A - \{j\}$. So $S_j(A - \{1\}) = A - \{1\}$, which means $A - \{1\}$ is in $S_j(\partial \mathcal{F})$. Therefore, all of ∂A is in $S_j(\partial \mathcal{F})$. Finally we assume that $j \in A$, $1 \notin A$ and $(A - \{j\}) \cup \{1\} \in \mathcal{F}$. Let $A - \{k\} \in \partial A$, $k \neq j$. Since $(A - \{k, j\}) \cup \{1\} \in \partial \mathcal{F}$, the switching map S_j fixes $A - \{k\}$, so $A - \{k\} = S_j(A - \{k\}) \in S_j(\partial \mathcal{F})$. The last case is $A - \{j\} = S_j(A - \{j\}) \in S_j(\partial A) \subset S_j(\partial \mathcal{F})$.

If $S_j(A) \neq A$, then $j \in A$, $1 \notin A$ and $S_j(A) = (A - \{j\}) \cup \{1\} \notin \mathcal{F}$. The subsets $(A - \{k, j\}) \cup \{1\}$ and $A - \{j\}$ in $S_j(\partial A)$ can be checked as in the previous paragraph.

Suppose we iterate the switching maps S_j for various j, until $\mathcal{F}$ is converted to a collection $\mathcal{F}'$ such that $S_j(\mathcal{F}') = \mathcal{F}'$ for all j. This is possible because S_j either fixes $\mathcal{F}$, or gives $\mathcal{F}$ more members which contain 1. Since the switching maps do not change the size of the collection, Lemma 4.5 implies that $|\partial \mathcal{F}'| \leq |\partial \mathcal{F}|$. This completes Step (1).

Step (2)   Let $\mathcal{F}' = \mathcal{F}'(1) \cup \mathcal{F}'(\tilde{1})$ as indicated. Clearly $\partial \mathcal{F}' = \partial \mathcal{F}'(1) \cup \partial \mathcal{F}'(\tilde{1})$. First we show that

(4.7) $\partial \mathcal{F}'(\tilde{1}) \subset \partial_1 \mathcal{F}'(1)$,

where ∂_1 is the operation of deleting 1 from a set. Let $B = A - \{j\} \in \partial \mathcal{F}'(\tilde{1})$, $j \neq 1$, $A \in \mathcal{F}'(\tilde{1})$. Since $\mathcal{F}'$ is fixed under S_j, $B \cup \{1\} \in \mathcal{F}'(1)$, so $B \in \partial_1 \mathcal{F}'(1)$. This establishes (4.7), which clearly implies

(4.8) $\partial \mathcal{F}' = \partial \mathcal{F}'(1)$.

We now separate the collection $\partial \mathcal{F}'$ into two subcollections: those subsets with 1 deleted, $\partial_1 \mathcal{F}'(1)$, and those subsets with some element $\neq 1$ deleted, $\partial(\partial_1 \mathcal{F}'(1)) \cup \{1\}$. (This notation means that we first delete 1 from each member of $\mathcal{F}'(1)$, next apply ∂, and then reinsert 1 into each member.) Because these two subcollections are disjoint,

(4.9) $|\partial \mathcal{F}'| = |\partial_1 \mathcal{F}'(1)| + |\partial(\partial_1 \mathcal{F}'(1)) \cup \{1\}|$.

This completes Step (2).

Step (3)   Suppose that

(4.10) $\quad |\partial_1 \mathcal{F}'(1)| \;\geq\; \binom{a_k-2}{k-1} + \dots + \binom{a_i-2}{i-1}.$

Because $\partial_1 \mathcal{F}'(1)$ is a collection of $(k-1)$-element subsets of the $(n-1)$-element subset $U-\{1\}$, we may conclude by induction that

(4.11) $\quad |\partial\partial_1 \mathcal{F}'(1)| \;\geq\; \binom{a_k-2}{k-2} + \dots + \binom{a_i-2}{i-2}.$

The insertion of 1 into each set of $\partial\partial_1\mathcal{F}'(1)$ is a bijection to $\partial(\partial_1\mathcal{F}'(1)) \cup \{1\}$. Thus, (4.9), (4.10) and (4.11) imply

$$|\partial \mathcal{F}| \;\geq\; \left[\binom{a_k-2}{k-1} + \binom{a_k-2}{k-2}\right] + \dots + \left[\binom{a_i-2}{i-1} + \binom{a_i-2}{i-2}\right]$$

which by Pascal's Triangle (§1.2, Eq. (2.3)) implies our result

(4.12) $\quad |\partial \mathcal{F}| \;\geq\; \binom{a_k-1}{k-1} + \dots + \binom{a_i-1}{i-1}.$

So we suppose that (4.10) is not true, i. e.,

(4.13) $\quad |\partial_1 \mathcal{F}'(1)| \;<\; \binom{a_k-2}{k-1} + \dots + \binom{a_i-2}{i-1}.$

Clearly $|\mathcal{F}| = |\mathcal{F}'| = |\mathcal{F}'(1)| + |\mathcal{F}'(\tilde{1})| = |\partial_1\mathcal{F}'(1)| + |\mathcal{F}'(\tilde{1})|$, so that (4.13) and the assumed value of $|\mathcal{F}|$ give

$$|\mathcal{F}'(\tilde{1})| \;>\; \left[\binom{a_k-1}{k} - \binom{a_k-2}{k-1}\right] + \dots + \left[\binom{a_i-1}{i} - \binom{a_i-2}{i-1}\right]$$

and again Pascal's Triangle implies

(4.14) $\quad |\mathcal{F}'(\tilde{1})| \;>\; \binom{a_k-2}{k} + \dots + \binom{a_i-2}{i}.$

This time $\mathcal{F}'(\tilde{1})$ is a collection of k-element subsets of the $(n-1)$-element set $U-\{1\}$, so again by the induction hypothesis

(4.15) $\quad |\partial\mathcal{F}'(\tilde{1})| \;>\; \binom{a_k-2}{k-1} + \dots + \binom{a_i-2}{i-1}.$

However (4.7) implies

(4.16) $\qquad |\partial_1 \mathcal{F}'(1)| \geq |\partial \mathcal{F}'(\tilde{1})|$.

Taken together, (4.15) and (4.16) contradict (4.13). So (4.10) must hold, and the proof is complete.

Notes

Three good general references for posets are [Ai], [Be] and [Gre-K1] . Sperner theorems are included in §3 of Chapter VIII of [Ai]. They are also a central topic of [Gre-K1]. Exercise 7 below, and Exercises 18 of Chapter 3 and Exercise 20 of Chapter 4 establish the entries of the chart for $\mathcal{L}_n$. The entries for $\mathcal{B}_n$ and $\mathcal{Y}_\lambda$ are given in Exercises 8 and 10. For $\mathcal{P}_n$ they are Exercises 9 and 22, for $\mathcal{D}_n$ Exercises 11 and 12, and for $\mathcal{C}_{nm}$ Exercises 13 and 29. A matching between two adjacent levels of the Boolean algebra can easily be shown to exist from Hall's Theorem. The decomposition into symmetric chains is not guaranteed from this technique. The fact that matching to first available in the Boolean algebra works is due to Aigner [Ai]. The relationship between the various matching schemes in $\mathcal{B}_n$ can be found in [Wh-Wi]. Kleitman's solution of the Littlewood-Offord problem appears in [Gre-K1]. The proof of the Kruskal-Katona Theorem is due to Frankl, [Fr].

Exercises

1.[2] Given a finite poset $(P, \leq)$, show that there is at least one way to totally order P, that is, label all of the elements of P with [n]: $a_1, \ldots, a_n$ so that $a_i \leq a_j$ implies $i \leq j$. Such a labeling is called a *linear extension* of $(P, \leq)$.

2.[2C] Write a program which takes as input a poset P and gives as output a linear extension of $(P, \leq)$.

3.[3C] For each poset $(P, \leq)$ pictured below, write a program which finds the total number of linear extensions of $(P, \leq)$. Can you prove any theorems here?

(a)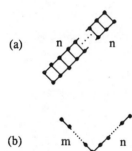

n   n

(b)   m   V   n

4.[2]   Let $(P, \leq)$ be a poset with a linear extension $a_1, \ldots, a_n$. The *incidence matrix* of $(P, \leq)$ is the $n \times n$ matrix Z, where $Z_{ij} = 1$ if $a_i \leq a_j$ and $Z_{ij} = 0$ otherwise. Prove that Z is invertible. Construct Z for B_3, P_3, $\mathcal{l}_3$ and $\mathcal{y}_{32}$.

5.[1]   For a ranked poset, the numbers $|L_i| = W_i$ are called the *Whitney numbers of the second kind*. For B_n, the Whitney numbers are the binomial coefficients. What are they for P_n? *showing us*

6.[2]   Let $W_k(n)$ denote W_k for $\mathcal{l}_n$. Prove
$$W_k(n) = W_k(n-1) + \ldots + W_{k-t}(n-1)$$
where $t = \min\{k, n-1\}$.

7.[1]   Prove that $\mathcal{l}_n$ is rank symmetric and order symmetric. (Exercise 18 of Chapter 3 will show that $\mathcal{l}_n$ is rank unimodal, and Exercise 20 of Chapter 4 shows that it is a lattice.)

8.[2]   Prove that B_n is rank unimodal, rank symmetric, order symmetric and a lattice.

9.[2]   Prove that P_n is a lattice.

10.[2]   Prove that $\mathcal{y}_\lambda$ is a lattice.

11.[2]   Prove that $\mathcal{D}_n$ is order symmetric.

12.[3]　Prove that $\mathfrak{D}_n$ is a lattice by finding the meet of two partitions λ and μ. (Hint: Take the smaller of the two partial sums to define the partial sum of the meet. Reason similarly for the join.)

13.[2]　Prove that C_{nm} is rank unimodal, rank symmetric and order symmetric. (The Sperner property follows from Exercise 29.)

✓ 14.[1]　What is the rank of $(n, n-1, \ldots, 1)$ in $\mathfrak{L}_n$?

15.[4C]　A chain C is called a *maximal chain* if C is not contained in any other chain. What is the length of a maximal chain in $\mathfrak{L}_{n+1}$? In $\mathfrak{Y}_\lambda$, $\lambda = (n, n-1, \ldots, 1)$? Write a program which gives the number of maximal chains each poset has. What are your conclusions? [Sta2]

✓ 16.[2]　How many maximal chains do B_n and P_n have?

17.[2]　How many edges do the Hasse diagrams of B_n and $\mathfrak{L}_n$ have?

✓ 18.[3C]　Write a program to count the number of edges in the Hasse diagram of P_n. Formulate a conjecture, based upon an appropriate combination of Bell numbers. (Hint: Try $(B_{n+2} - a \cdot B_{n+1} + b \cdot B_n) / 2$, for appropriate positive integers a and b. Can you give a combinatorial proof of this result?)

19.[4C]　Write a program to compute the Whitney numbers of $\mathfrak{Y}_\lambda$, $\lambda = n^m$. Formulate and prove as many conjectures as you can. §3.3 may be useful.

20.[2]　Suppose $v_0, v_1, \ldots, v_n$ is a finite sequence of positive real numbers. We say $v_0, v_1, \ldots, v_n$ is *log-concave* if $v_k^2 \geq v_{k-1} \cdot v_{k+1}$ for $1 \leq k \leq n-1$. If $v_0, v_1, \ldots, v_n$ is log-concave, prove that it is also unimodal.

21.[3]　Suppose the polynomial $v(x) = v_n + v_{n-1} x^1 + \ldots + v_0 x^n$ has negative real roots. Show that $v_0, v_1, \ldots, v_n$ is log-concave. Possible hint: This result has a combinatorial proof. Let $\{-r_1, \ldots, -r_n\}$ be the roots of $v(x)$. Define the weight $w(A)$ of a subset $A \subset [n]$ to be

$$w(A) = \prod_{i \in A} r_i.$$

Show that

$$v_k^2 = \sum_{\substack{(A,B) \\ |A|=|B|=n-k}} w(A)w(B),$$

and

$$v_{k-1}\,v_{k+1} = \sum_{\substack{(C,D) \\ |C| = n-k+1 \\ |D| = n-k-1}} w(C)w(D).$$

Then use the unimodality of the binomial coefficients to give an injection of the pairs (C,D) for $v_{k-1}v_{k+1}$ to (A, B) for v_k^2 which preserves the weight.

22.[3] Let $S_n(x)$ be the polynomial of degree n

$$S_n(x) = \sum_{k=1}^{n+1} S(n+1, k)\, x^{k-1}.$$

(a) Using (5.1) of Chapter 1, show that

$$S_n(x) = (x + 1)\, S_{n-1}(x) + x\, S'_{n-1}(x).$$

(b) Now show that $S_n(x)$ has n distinct negative roots following these steps. By induction, suppose that $S_{n-1}(x)$ has $n-1$ distinct negative roots $-r_1 < \ldots < -r_{n-1}$. From (a), show that $S_n(-r_i)$ and $S_n(-r_{i+1})$ have opposite signs for each i, $1 \le i \le n-2$. Conclude that $S_n(x)$ must have a root in the interval $(-r_i, -r_{i+1})$, so that $S_n(x)$ has at least $n-2$ negative roots. Show that $S'_{n-1}(-r_{n-1}) > 0$, and again use (a) to show that $S_n(-r_{n-1}) < 0$. Conclude that $S_n(x)$ has a root in the interval $(-r_{n-1}, 0)$. Finally, conclude that $S_n(x)$ has another root in the interval $(-\infty, -r_1)$. We have just shown that the roots of $S_n(x)$ and $S_{n-1}(x)$ *interlace*: between any two consecutive roots of one polynomial there is exactly one root of the other polynomial.

(c) From Exercises 20 and 21 show that the Stirling numbers of the second kind are unimodal.

23.[2]   Let S be a subset of $[999] \cup \{0\}$. If $|S| \geq 76$, show that S must contain at least two numbers n and m, so that the difference $n - m$ can be computed with no "borrowing". (Hint: consider the poset $C_{3,10}$.)

24.[3]   Let $A = \{a_1, \ldots, a_p\} \subset [n]$, where $a_1 < \ldots < a_p$. Define $g(A) = \{a_1, \ldots, a_t, a_t + 1, a_{t+1}, \ldots, a_p\}$, where t is the largest i for which $a_i - 2i$ is minimal (assume $a_0 = 0$). Prove that $g(A) = f(A)$, where f is defined in Algorithm 14. [Ai]

25.[3]   Suppose $A \subset [n]$ is represented as a sequence of n parentheses, where parenthesis i is right if $i \in A$ and left otherwise. Thus,

$$A = \{1, 3, 4, 6, 7, 10, 12, 13, 16, 19\} \subset [21]$$

corresponds to

```
)  (  )  )  (  )  )  (  (  )  (  )  )  (  (  )  (  (  )  (  (
1  2  3  4  5  6  7  8  9 10 11 12 13 14 15 16 17 18 19 20 21
```

and if we pair off the parentheses in the usual way,

```
)  (  )  )  (  )  )  (  (  )  (  )  )  (  (  )  (  (  )  (  (
1  2  3  4  5  6  7  8  9 10 11 12 13 14 15 16 17 18 19 20 21
```

we are left with a string of unpaired right parentheses followed by a string of unpaired left parentheses.

```
)  )  )  (  (  (  (
1  4  7 14 17 20 21
```

Now take the first unpaired left parenthesis and turn it around.

```
)  )  )  )  (  (  (
1  4  7 [14] 17 20 21
```

The resulting string of parentheses

```
)  (  )  )  (  )  )  (  (  )  (  )  )  (  )  (  (  )  (  (
1  2  3  4  5  6  7  8  9 10 11 12 13 14 15 16 17 18 19 20 21
```

corresponds to a new subset

$$h(A) = \{1, 3, 4, 6, 7, 10, 12, 13, 14, 16, 19\}.$$

a) Prove that h satisfies the conclusion of Theorem 2.3.

b) Prove that h gives a symmetric chain decomposition of B_n.

c) Prove that h = f. ([Gre-K2]; see also [Ai], p. 439.)

26.[4C]   Write a program which applies Algorithm 14 to colex order. Try it for various values of n and p. Does it work? What can you conjecture (and prove) about the resulting f? [Wh-Wi]

27.[2C]   Use the first available match to find a matching for P_n for some small values of n. You can use the lex order of RG functions given in Algorithm 11 instead of the lex order of subsets of [n]. What is your chain decomposition?

28.[2]   Let $m = p_1 p_2 \cdots p_n$, where each p_i is a different prime. Show that the maximum number of divisors of m which do not divide each other is M(n).

29.[2]   Find a symmetric chain decomposition for C_{nm} (p. 436 of [Ai]).

30.[4C]   Suppose the integer partitions of n are ordered by reverse refinement. That is, $\lambda <\cdot \mu$ if and only if they are identical except one part of λ is split into two parts of μ. For instance, $5^1 \, 3^2 \, 1^1 <\cdot 4^1 \, 3^2 \, 1^2$.

   (a) Show that the resulting poset is ranked with a $\hat{0}$ and a $\hat{1}$, but it is not rank or order symmetric.

   (b) Show that the resulting poset is not a lattice.

   (c) Revise Algorithms 7 and 14 to investigate the unimodal and Sperner properties. What theorems can you guess? What theorems can you prove?

31.[3]   Let L_k and L_{k-1} be two levels in B_n with $k \leq \lfloor n/2 \rfloor$. Suppose T is an independent set from these two levels. Prove that $|T|$ is maximal if and only if $T = L_k$.

32.[4]   Generalize the Erdös-Ko-Rado Theorem to collections $\mathcal{F}$ of subsets of [n] such that $A, B \in \mathcal{F}$ implies $|A| \le k$ and $A \cap B \ne \varnothing$.

CHAPTER 3

Bijections

We have already encountered several examples of explicit bijections $\varphi : A \rightarrow B$, for two finite sets A and B. In Chapter 1 we let A be the set of all permutations π of $[n]$ and B be the set $\{0, 1, \ldots, n! - 1\}$. The rank function was an explicit bijection from A to B. It was closely related to the listing algorithm for permutations.

There are many reasons for constructing a bijection $\varphi : A \rightarrow B$. The most obvious application is to conclude that $|A| = |B|$. This is useful if we know $|A|$ but do not know $|B|$. In some cases A might have a complicated subset A_0 which interests us. The bijection could simplify A_0 to $B_0 = \varphi(A_0)$. It might be the case that A and B are both complicated; we can still conclude that $|A| = |B|$. But in any case, bijections can be used to explain why certain classes of objects are counted by the same number.

Bijections can also be used to establish generating functions. Let's take a very simple example, the binomial theorem, to show this. Let A be the set of all subsets of $[n]$. The weight $w(\alpha)$ of an element of A is defined by

(0.1) $\qquad w(\alpha) = x^{|\alpha|}.$

So the generating function of all elements of A is

(0.2) $\qquad f(x) \; = \; \sum_{\alpha \in A} x^{|\alpha|} \; = \; \sum_{\alpha \in A} w(\alpha).$

We know that A has

$$\binom{n}{k}$$

elements α such that $|\alpha| = k$, so

$$(0.3) \qquad f(x) = \sum_{k=0}^{n} \binom{n}{k} x^k.$$

Let B be the set of n-tuples of 0's and 1's, and let the weight of any element β of B be $w(\beta) = x^k$, where β has k 1's. Define the bijection $\varphi : A \to B$ by

$$\varphi(\alpha)(i) = \begin{cases} 1 & \text{if } i \in \alpha \\ 0 & \text{if } i \notin \alpha. \end{cases}$$

The bijection φ is *weight-preserving*, because $w(\alpha) = w(\varphi(\alpha))$. It is clear that the generating function for B is $(1 + x)^n$, so (0.2) becomes

$$(0.4) \qquad f(x) = (1 + x)^n.$$

Another kind of combinatorial proof of the binomial theorem can be given when x is a non-negative integer. In this case, the right-hand side of (0.4) counts functions from $[n]$ to a set S of size $1 + x$. The right-hand side of (0.3) counts the same functions by classifying by the number of members of $[n]$ which get sent to the first x members of S. The theorem can easily be extended to all real x.

In this chapter we give examples of all these phenomena. The Catalan numbers provide bijections between apparently unrelated sets. The Prüfer correspondence associates a simple B with a complicated A. Partitions and permutations illustrate all these ideas. The most difficult bijection that we consider is the Schensted correspondence between permutations and tableaux.

Several of the constructions in this chapter and the next involve graphs. While none of the graph theory concepts used are difficult, we will state here some of the key definitions and results involving graphs.

A *graph* is a set of *vertices* and a set of *edges*. The edges are usually a collection of 2-element subsets of the vertex set (such graphs are called *simple*), but sometimes one-element subsets (*loops*) or repetitions (*multiple edges*) are allowed. If the edges are ordered pairs of vertices, the graph is called *directed* (or a *digraph*) and the edges are *directed edges*.

The *degree* of a vertex is the number of vertices *incident* to it. A vertex in a directed graph has an *in-degree* and an *out-degree*.

A *path* in a graph is a sequence of *adjacent* vertices. A path is *simple* if no vertex is repeated (except that the first may equal the last). A *cycle* is a simple path which starts and ends at the same vertex. We sometimes will refer to the unordered set of vertices in a cycle as a cycle. A graph is *connected* if there is a path from every vertex to every other vertex. Directed graphs have *directed paths*. Directed graphs are *strongly connected* if there is a directed path from every vertex to every other vertex. Graphs may have one *component* (connected) or several components.

A *tree* is a connected simple graph with no cycles. A tree on n vertices has n − 1 edges. There is a unique path between any pair of vertices in a tree. Trees may be *rooted*, i. e., have a distinguished vertex. In that case, the tree may be directed naturally along the unique path from any vertex to the root. In rooted trees, vertices are related in a familial sense: they may be *fathers*, *brothers* or *sons*.

A *bipartite graph* is a graph whose vertex set is partitioned into two blocks. All edges in the graph go between these two blocks. A *complete graph* is the simple graph where every possible edge is drawn.

Graphs may be *labeled* or *unlabeled*. Generally speaking, labeled graphs are easier to deal with. For example, there are

$$2^{\binom{n}{2}}$$

labeled simple graphs on n vertices, but the number of unlabeled graphs is given by a complicated formula which involves Polya's enumeration theorem.

§3.1 The Catalan Family

There is a sequence of integers, called the *Catalan numbers*, which occur frequently in combinatorial problems. They are defined by

$$(1.1) \qquad C_n = \frac{1}{n+1} \binom{2n}{n},$$

so that $C_0 = 1$, $C_1 = 1$, $C_2 = 2$, $C_3 = 5$, $C_4 = 14$, etc. Many combinatorial objects are counted by these numbers.

We will describe bijections between six sets and then show that one of these sets is counted by the Catalan numbers. Three of these sets will involve trees. These sets are:

(1) Binary trees A *binary tree* is a rooted tree where each vertex has either 0, 1 or 2 sons; and, when only one son is present, it is either a *right son* or a *left son*.

(2) Ordered trees An *ordered tree* can be defined inductively as a rooted tree whose principal subtrees (the trees obtained by removing the root) are ordered trees and have been assigned some order among themselves.

(3) Full binary tree A *full binary tree* is a binary tree where every vertex has either 0 or 2 sons.

(4) Well-formed parentheses A sequence of parentheses is called *well-formed* if, at any point in the sequence, the number of right parentheses up to this point does not exceed the number of left parentheses up to the same point. Moreover, the total number of left parentheses equals the total number of right parentheses.

(5) Ballot problem Suppose Alice and Barbara are candidates for office. The result is a tie. In how many ways can the ballots be counted so that Alice is always ahead of or tied with Barbara?

(6) Standard tableaux Given a partition λ of n, a *standard tableau* T is an arrangement of [n] in the n cells of the Ferrers diagram of λ which increase across rows and down columns. These objects will be discussed in greater detail in §3.5.

THEOREM 1.1 *The following sets of objects all have the same number of elements, and this number is* C_n:

(1) *binary trees on* n *vertices*;
(2) *ordered trees on* $n + 1$ *vertices*;
(3) *full binary trees on* $2n + 1$ *vertices*;
(4) *well-formed sequences of* 2n *parentheses*;
(5) *solutions to the ballot problem when* 2n *votes are cast; and*
(6) *standard tableaux in a* $2 \times n$ *rectangular Ferrers diagram*.

Proof We use bijections to show (1)-(6) are equinumerous. Then we show that (4) yields the Catalan number.

(1) = (2): We give a bijection φ from binary trees to ordered trees. Let B be a binary tree. Here is how we construct $T = \varphi(B)$.

(a) The vertices of B are the vertices of T with the root deleted.
(b) The root of B is the first son of the root of T.
(c) Vertex v is a left son of vertex w in B if and only if v is the first son of w in T.
(d) Vertex v is a right son of vertex w in B if and only if v is the brother to the right of w in T.

In the examples below, we have labeled the vertices to help the reader trace what happens under φ.

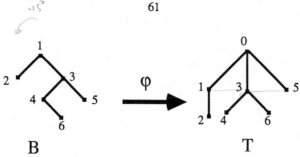

B   φ   T

(1) = (3): The bijection φ will be from binary trees to full binary trees. Let B be a binary tree. Construct $F = \varphi(B)$ by adding a new son to each vertex of B with exactly one son, and adding two sons to each vertex of B with no sons (terminal vertices).

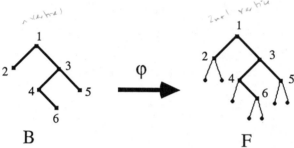

B   φ   F

You are asked to prove in Exercise 2 that the number of terminal vertices in a full binary tree is one more than the number of internal (non-terminal) vertices. So F is a full binary tree on $2n + 1$ vertices. It is clear that φ^{-1} is given by "pruning" the terminal vertices from F.

(2) = (4): The bijection φ will be from ordered trees to well-formed sequences of parentheses. Let T be an ordered tree. We show how to construct $P = \varphi(T)$. If T consists of a single vertex (the root), then P is the empty sequence. Now φ will be defined recursively. Suppose φ has been defined for all ordered trees $\tilde{T}$ with $k + 1$ vertices, $k < n$, and $\varphi(\tilde{T})$ has $2k$ parentheses. Let T be an ordered tree with $n + 1$ vertices and principal subtrees $T_1, T_2, \ldots, T_s$. Let $P_1, P_2, \ldots, P_s$ be the corresponding well-formed sequences of parentheses. Then $P = (P_1)(P_2) \ldots (P_s)$. Clearly, the number of parentheses in P is

$$2s + \sum_{i=1}^{s} 2(\# \text{ of vertices in } T_i - 1) = 2n.$$

It is also clear that P is well-formed.

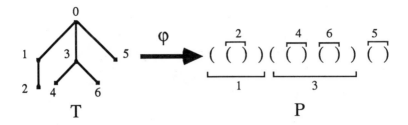

To define φ^{-1}, let $\varphi^{-1}(\varnothing)$ = the tree consisting of only a root. Again, by induction, assume $\varphi^{-1}(\tilde{P}) = \tilde{T}$ has been defined for all sequences $\tilde{P}$ of length $2k$ and ordered trees $\tilde{T}$ with $k+1$ vertices, $k < n$. Let P be a well-formed sequence of length $2n$. Write $P = \tilde{P}_1 \tilde{P}_2 \dots \tilde{P}_s$, where each $\tilde{P}_i$ is determined by those points in P where the number of left and right parentheses are equal. Each $\tilde{P}_i$ is itself a well-formed sequence enclosed in a parenthesis pair. Let P_i be the well-formed sequence obtained from $\tilde{P}_i$ by removing this pair. Let T_i be the associated tree. Put $\varphi^{-1}(P) = T$, where the ordered tree T has principal subtrees $T_1, \dots, T_s$.

(4) = (5): Now the bijection φ will be from well-formed parentheses sequences to solutions to the ballot problem. Let P be a well-formed sequence. Then $W = \varphi(P)$ is obtained by replacing "(" by a vote for Alice (or "A") and ")" by a vote for Barbara (or "B"). The word W in the letters "A" and "B" is a sequence of votes which solves the ballot problem since P is well-formed.

$$(\,(\,)\,)\,(\,(\,)\,(\,)\,)\,(\,) \xrightarrow{\;\varphi\;} \text{AABBAABABBAB}$$
$$\qquad\qquad P \qquad\qquad\qquad\qquad W$$

(4) = (6): The bijection φ will be from well-formed sequences to standard tableaux of shape $2 \times n$. Let P be a well-formed sequence; define $S = \varphi(P)$ as follows. Label the positions of the $2n$ parentheses in P by $1, 2, \dots, 2n$. In S, put i in the first row if and only if the parenthesis in position i is left. Otherwise, put i in the second row. Arrange the entries of S to be increasing in the two rows. Since P is well-formed, S must be increasing down the columns. The definition of φ^{-1} is clear.

$$\begin{array}{cccccccccccc} 1 & 2 & 3 & 4 & 5 & 6 & 7 & 8 & 9 & 10 & 11 & 12 \\ (\, & (\, &)\, &)\, & (\, & (\, &)\, & (\, &)\, &)\, & (\, &)\, \end{array} \xrightarrow{\;\varphi\;}$$

1	2	5	6	8	11
3	4	7	9	10	12

$$\qquad\qquad\qquad P \qquad\qquad\qquad\qquad\qquad\qquad S$$

Finally, we show that the number of sequences of well-formed parentheses of length $2n$ is C_n. To begin with, rewrite C_n as follows:

(1.2) $C_n = \dfrac{1}{2n+1}\dbinom{2n+1}{n+1}.$

Let us represent sequences of parentheses as strings of 0's and 1's (called *bit strings*): 0 to represent a right parenthesis and 1 a left parenthesis. Such a sequence can also be represented as a lattice path, such as those described in §2.2. In view of (1.2), we consider $\mathscr{S}_n$, the set of bit strings with $2n+1$ digits, $n+1$ of which are 1's. Notice that a cyclic permutation of any bit string in $\mathscr{S}_n$ gives a new bit string in $\mathscr{S}_n$. In fact, a bit string in $\mathscr{S}_n$ can be cyclically permuted into $2n+1$ distinct bit strings. Thus $\mathscr{S}_n$ can be partitioned into equivalence classes of size $2n+1$. The number of such classes is exactly C_n.

Now let us consider one of these equivalence classes, $\mathscr{O}$. Pick any bit string $w \in \mathscr{O}$. Label the positions between the digits of w by $1, 2, \dots, 2n+2$. Let $\gamma(w)$ be the position of the <u>rightmost minimum</u> in the lattice path of this bit string.

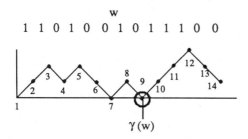

Let w^+ be the same bit string cyclically permuted one position to the left:

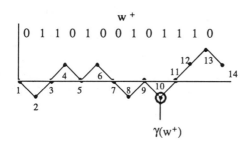

Observe that

$$\gamma(w^+) = \begin{cases} \gamma(w) + 1 & \text{if } \gamma(w) \neq 2n+1 \\ 1 & \text{if } \gamma(w) = 2n+1. \end{cases}$$

Thus, there is exactly one string $w^* \in \mathcal{O}$ such that $\gamma(w^*) = 1$.

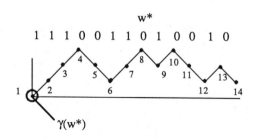

w*

1 1 1 0 0 1 1 0 1 0 0 1 0

Note that the initial digit in w^* must be 1. If this 1 is removed, what remains is the bit string corresponding to a well-formed sequence of parentheses. Conversely, if a 1 is added to the front of the bit string of a well-formed sequence to form the bit string $\hat{w}$, then $\gamma(\hat{w}) = 1$.

This means that the number of well-formed sequences of 2n parentheses is exactly the number of equivalence classes, i. e., C_n.

∎

Several more examples of sets counted by Catalan numbers are given in the exercises.

§3.2 The Prüfer Correspondence

A bijection $\varphi : A \to B$ can be used to transfer properties of A to a simpler set B. The Prüfer correspondence provides an example of this phenomenon. In this case, A consists of all labeled trees on n vertices, while B is just the set of all (n−2)-tuples of integers in [n].

Recall that a tree is a connected graph with no cycles and a labeled tree on n vertices is just a tree whose n vertices are labeled with the integers in [n]. There are clearly

$$2^{\binom{n}{2}}$$

labeled graphs on n vertices. We wish to find the number of labeled trees on n vertices. The Prüfer correspondence φ establishes Cayley's Theorem.

THEOREM 2.1 *The number of labeled trees on* n *vertices is* n^{n-2}.

Given a labeled tree T on n vertices, we want to produce $\varphi(T)$, an (n–2)-tuple with entries in [n]. The construction of $\varphi(T)$ involves removal of terminal vertices from T. Terminal vertices always exist in view of the next lemma.

LEMMA 2.2 *Every tree* T *with two or more vertices has two or more terminal vertices.*

Proof A well-known property of trees on n vertices is that they have n − 1 edges. Since T has n − 1 edges, the sum of the degrees of all vertices is $2 \cdot (n - 1)$. Every degree is at least 1, so the pigeonhole principle implies that at least two degrees are exactly 1.

Now $\varphi(T) = (a_1, \ldots, a_n)$ is easily described. Let $\mathfrak{I}$ be the set of terminal vertices of T, so $\mathfrak{I} \subset [n]$. Let $v = \max\{i : i \in \mathfrak{I}\}$ and a_1 be the vertex adjacent to v. Now delete v and the edge $v - a_1$ from T to form a new tree $\tilde{T}$ and iterate. The iteration stops when the tree has only two vertices remaining. We will then have an (n–2)-tuple, $\varphi(T) = (a_1, \ldots, a_n)$. Here is an example of this construction.

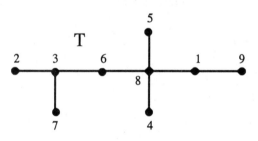

$$\varphi(T) = (1, 3, 8, 8, 3, 6, 8)$$

In the following description of this procedure, [n] is the vertex set of T and E is the edge set. The function adj(v) will return the vertex adjacent to $v \in \mathfrak{I}$.

ALGORITHM 15: *The Prüfer Correspondence*

begin

    for $k \leftarrow 1$ **to** $n - 2$ **do**

        $v \leftarrow \max\{i : i \in \mathfrak{I}\}$

$$a_k \leftarrow adj(v)$$
$$E \leftarrow E - \{v — a_k\}$$

end.

To show φ is a bijection, we describe φ^{-1}. Let $(a_1, \ldots, a_n)$ be an (n–2)-tuple with entries in [n]. If T satisfies $\varphi(T) = (a_1, \ldots, a_n)$, then the degree of i in T is one more than the number of occurrences of i in $\varphi(T)$. In fact, the terminal vertices $\mathfrak{I}$ are precisely those which do not occur in $\varphi(T)$. Let $v = \max\{i : i \in \mathfrak{I}\}$ and construct the edge $a_1 — v$. The new terminal vertices are found by subtracting one from the degree of a_1 and removing v from $\mathfrak{I}$. In this way $n-2$ edges are constructed. The final edge is drawn between the only remaining terminal vertices, which will be a_{n-2} and 1 (or 2 if $a_{n-2} = 1$). It is clear that the edges of T are inserted in the same order that φ removed them. We leave the details of the induction to the reader.

In the algorithm below, the vector $(d_1, \ldots, d_n)$ keeps track of the residual degrees of the vertices of T.

ALGORITHM 16: *The Prüfer Correspondence*

begin

    for $v \leftarrow 1$ **to** n **do**

        $d_v \leftarrow 1$

    for $i \leftarrow 1$ **to** n - 2 **do**

        $v \leftarrow a_i$

        $d_v \leftarrow d_v + 1$

    $E \leftarrow \varnothing$

    for $i \leftarrow 1$ **to** $n - 2$ **do**

        $w \leftarrow \max\{k : d_k = 1\}$

        $v \leftarrow a_i$

        $E \leftarrow E \cup \{w — v\}$

        $d_v \leftarrow d_v - 1$

        $d_w \leftarrow 0$

    $v \leftarrow \max\{k : d_k = 1\}$

    $w \leftarrow \min\{k : d_k = 1\}$

    $E \leftarrow E \cup \{w — v\}$

end.

The following theorem is typical of the kinds of results that can be obtained from the Prüfer correspondence.

THEOREM 2.3 *The number of labeled trees with* t *terminal vertices is*

$$(n! \, / \, t!) \cdot S(n-2, n-t).$$

Proof The number of labeled trees with t terminal vertices is

$$\binom{n}{t}$$

times the number of labeled trees with $\{1, 2, \ldots, t\}$ as terminal vertices. The Prüfer code $\varphi(T)$ for such a tree does not include any entries from $\{1, 2, \ldots, t\}$. Moreover, $\{t+1, \ldots, n\}$ each must occur at least once in $\varphi(T)$ since they are not terminal vertices. Let $B_{t+1}, \ldots, B_n$ be the sets of positions where $t+1, \ldots, n$ appear in $\varphi(T)$. These blocks define a partition of $[n-2]$ into $n-t$ labeled blocks. The number of such set partitions (see §1.5) is $S(n-2, n-t)$, and the number of ways of labeling the blocks is $(n-t)!$. So the number of labeled trees with t terminal vertices is

$$\binom{n}{t} (n-t)! \; S(n-2, n-t) \; = \; \frac{n!}{t!} \; S(n-2, n-t).$$

There is a startling application of Theorem 2.1 to permutations. Suppose π is an *n-cycle*, that is, a permutation whose cycle decomposition consists of one cycle of length n. We may ask in how many ways can π be written as a product of $n-1$ transpositions? For example, if $\pi = (123)$, then $\pi = (23) \circ (13)$, $\pi = (13) \circ (12)$ or $\pi = (12) \circ (23)$, so there are three such representations. The answer is rather unexpected.

THEOREM 2.4 *Given an* n-cycle π, *the number of sequences* $(t_1, \ldots, t_{n-1})$ *of transpositions such that* $\pi = t_1 \circ t_2 \circ \ldots \circ t_{n-1}$ *is* n^{n-2}.

Proof We will count the total number N of sequences $(t_1, \ldots, t_{n-1})$ of transpositions whose product is an n-cycle. Since there are $(n-1)!$ n-cycles, the number of such sequences which will yield a <u>specific</u> n-cycle will then be

$N/(n-1)!$.

We shall need two graphs, one associated with a collection of transpositions and one associated with a permutation. Given a set of transpositions $\{t_1, \ldots, t_k\}$, let T be the labeled graph whose vertices are given by $[n]$ and whose edges are given by the transpositions $\{t_1, \ldots, t_k\}$. Given any permutation π, let G_π be the directed graph on $[n]$ with edges $i \to j$ if and only if $\pi(i) = j$. Notice that the fixed points of π correspond to directed loops in G_π and the cycle decomposition of π corresponds to the decomposition of G_π into directed cycles.

The following lemma shows how G_π is affected by multiplying by a transposition.

LEMMA 2.5 *Let* t *be the transposition* (x y) *and* $\tilde{\pi} = \pi \circ t$. *Then the graph of* $G_{\tilde{\pi}}$ *is obtained from* G_π *in the following way:*

(1) *If* x *and* y *are in different cycles of* G_π, *then these two cycles merge into one cycle;*

(2) *If* x *and* y *are in the same cycle of* G_π, *then this cycle splits into two cycles.*

Proof The proof is clear from the following two pictures. In Case (1) we have:

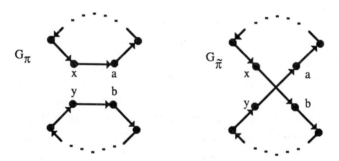

In Case (2) we have:

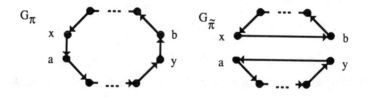

That $N = n^{n-2}(n-1)!$ follows immediately from this last lemma.

LEMMA 2.6 *Suppose* $\pi = t_1 \circ t_2 \circ \ldots \circ t_{n-1}$ *and* T *is the graph corresponding to* $\{t_1, \ldots, t_{n-1}\}$. *Then* π *is an n-cycle if and only if* T *is a tree.*

Proof First we suppose T is a tree and prove π is an n-cycle. The proof is by induction on n. The edge e corresponding to t_{n-1} is a *cut-edge* of T, i. e., its removal will disconnect T. So $T - e = T_1 \cup T_2$ for two smaller trees T_1 and T_2, where T_1 has k vertices and T_2 has $n - k$, $k \geq 1$. By induction, the product of the transpositions t_i for T_1 and T_2 (taken in the same order as in π) is a k-cycle σ_1 and an (n–k)-cycle σ_2 respectively. Moreover, since $T_1 \cap T_2 = \varnothing$, the transpositions for T_1 commute with those for T_2. Thus $t_1 \circ t_2 \circ \ldots \circ t_{n-2} = \sigma_1 \circ \sigma_2$ and the graph $G_\pi \circ t_{n-1}$ consists of these two cycles. Then G_π is obtained from $G_\pi \circ t_{n-1}$ by part (1) of Lemma 2.5.

Now suppose π is an n-cycle and we prove T is a tree. Consider what happens to the graphs $G_{\pi(k)}$, $\pi^{(k)} = t_1 \circ t_2 \circ \ldots \circ t_k$, as we successively add edges to construct T. Initially, $G_{\pi(0)}$ consists of n loops and T is empty. Since G_π has only one component and we are adding exactly $n - 1$ edges, we must be in case (1) of Lemma 2.5 for each new transposition t_{k+1}. So we assume by induction that $G_{\pi(k)}$ has $n - k$ cycles $\{C_i\}$, and T is a *forest* of $n - k$ trees $\{T_i\}$, with the vertices of T_i = the vertices of C_i. Then the transposition t_{k+1} combines the two cycles (C_1 and C_2) to form $\tilde{C}_1$ in $G_{\pi(k+1)}$ and connects trees T_1 and T_2 to form a new tree $\tilde{T}_1$. Clearly the vertices of $\tilde{T}_1$ = the vertices of $\tilde{C}_1$. Thus $G_{\pi(k+1)}$ has the desired properties. Putting $k = n - 1$ gives a forest of one tree, i. e., a tree. This completes the proof of Lemma 2.6 and also Theorem 2.4.

■

Unfortunately, this is not a <u>direct</u> proof of Theorem 2.4. Such a proof would establish a bijection between labeled trees and sequences of transpositions whose product is a given n-cycle.

§3.3 Partitions

Integer partitions provide a rich source of bijections. We already gave a simple bijection for partitions in Theorem 3.1 of Chapter 1. In this section are several other examples. The first theorem is due to Euler.

THEOREM 3.1 *The number of partitions of* n *into odd parts equals the number of partitions of* n *into distinct parts.*

Proof Let PO(n) and PD(n) denote the two sets of partitions. We will construct a bijection $\varphi : PO(n) \to PD(n)$. Let $\lambda \in PO(n)$ whose largest part is N and let i be an odd part of λ of multiplicity m_i. Write m_i in base 2:

$$m_i \; = \; \sum_j a_j \, 2^j, \quad a_j \; = \; 0 \text{ or } 1.$$

For each $a_j = 1$, φ will create a part of size $i \cdot 2^j$. We write $\varphi(m_i, i)$ to denote these parts. Clearly they are distinct. Now let $\varphi(\lambda) = \sum \varphi(m_i, i)$. Note that if $i_1 \cdot 2^{j_1} = i_2 \cdot 2^{j_2}$ with i_1 and i_2 both odd, then $i_1 = i_2$ and $j_1 = j_2$.

The definition of φ^{-1} is straightforward. Take all parts of $\lambda \in PD(n)$ of the form $i \cdot 2^j$, for some j and a fixed odd i. Then $\varphi^{-1}(\lambda)$ has

$$\sum_j 2^j$$

parts of size i.

As an example of this bijection, take $\lambda = 15^1 \, 9^2 \, 5^3 \, 3^3 \, 1^4$. The 4 in 1^4 written base 2 is 100, so $\varphi(4, 1) = 4$. The exponent 3 in 3^3 is 11 in binary, so $\varphi(3, 3) = 3 + 6$. Do the same thing for the other parts. Then put the parts together to get $\varphi(15^1 \, 9^2 \, 5^3 \, 3^3 \, 1^4) = (18, 15, 10, 6, 5, 4, 3)$.

As generating functions, Theorem 3.1 is clearly equivalent to

$$(3.1) \qquad \prod_{i=0}^{\infty} (1 - x^{2i+1})^{-1} \; = \; \prod_{n=0}^{\infty} (1 + x^{n+1}).$$

In fact, the bijection φ is really just a combinatorial proof of

$$(3.2) \qquad (1 - x^{2i+1})^{-1} \; = \; \prod_{j=0}^{\infty} (1 + x^{(2i+1)2^j})$$

which provides a proof of (3.1).

The next example is of a somewhat different nature. If λ is a partition whose largest part λ_1 is $\le n$, and whose number of parts is $\le m$, then the Ferrers diagram of λ is contained in an $m \times n$ rectangle. The next theorem counts these partitions.

THEOREM 3.2 *The number of partitions λ whose Ferrers diagram lies inside an $m \times n$ rectangle is*

$$\binom{n+m}{n}.$$

Proof We construct the bijection $\varphi : P_{mn} \to B$, where P_{mn} is the set of partitions λ and B is the set of (n+m)-tuples with m 0's and n 1's.

As usual, let the $m \times n$ rectangle be located in the 4th quadrant in the xy-plane. Then the outside border of λ is the set of $n + m$ line segments connecting $(0, -m)$ to $(n, 0)$ along λ. For example, if $\lambda = (3, 3, 1)$, $n = 5$ and $m = 4$, the outside border of λ is shown below.

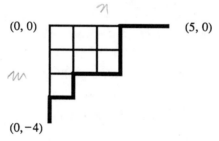

It is clear that λ is uniquely determined by this lattice path from $(0, -m)$ to $(n, 0)$, with unit steps up or to the right. The converse is also true. We code such a lattice path with an (n+m)-tuple of 0's and 1's. An up step is 0 and a step to the right is a 1. So we get m 0's and n 1's. This determines $\varphi(\lambda)$. In the example above, $\varphi(\lambda)$ = (010110011).

# of dots in Ferres diagram

For such a partition λ, let $\|\lambda\|$ denote the number λ partitions. The generating function

(3.3) $\qquad G_{mn}(q) \; = \; \sum_{\lambda \in P_{mn}} q^{\|\lambda\|} \qquad = q_0 + c_1 q^1 + c_2 q^2 \cdots \; c_{mn} q^{mn}$

$c_i = $ # of partitions of i inside $n \times m$ rect.

is a polynomial in q of degree m·n. According to Theorem 3.2,

if $q = 1$ then

$c_0 + c_1 + \cdots \; c_{mn} = \binom{n+m}{n}$

$$G_{mn}(1) = \binom{n+m}{n}.$$

Thus $G_{mn}(q)$ is a polynomial analogue of the binomial coefficient. It is called the *q-binomial coefficient* and is denoted

$$(3.4) \qquad G_{mn}(q) = \begin{bmatrix} n+m \\ n \end{bmatrix}_q.$$

It is clear that $G_{mn}(q)$ is the rank generating function of the Young lattice $\mathcal{Y}_\lambda$ of the partition $\lambda = n^m$. Thus properties of this lattice from Chapter 2 become properties of the q-binomial coefficient.

A partition λ is *self-conjugate* if $\lambda = \lambda'$ (see §1.3), i. e., the Ferrers diagram of λ is invariant under a flip across the main diagonal.

THEOREM 3.3 *The number of self-conjugate partitions of* n *is equal to the number of partitions of* n *into odd, distinct parts.*

Proof Let PSC(n) and POD(n) be the appropriate sets of partitions. We define $\varphi : \mathrm{PSC}(n) \to \mathrm{POD}(n)$. For $\lambda \in \mathrm{PSC}(n)$, $\lambda = (\lambda_1, \dots, \lambda_s)$, let d be the largest integer such that $\lambda_k \geq k$. In fact, d is the size of the largest principal subsquare (called the *Durfee square*) of the Ferrers diagram of λ. Or, d is the length of the main diagonal of the Ferrers diagram of λ. Number the cells down this diagonal $1, 2, \dots, d$. The cell numbered k will have $\lambda_k - k$ cells to its right, and since λ is self-conjugate, $\lambda_k - k$ cells below it. So the cells of λ are partitioned into blocks, one block for each diagonal cell, with the block corresponding to diagonal cell k containing $2\lambda_k - 2k + 1$ cells. Since $\lambda_1 \geq \dots \geq \lambda_d$ these odd numbers are distinct. Let $\varphi(\lambda) = (2\lambda_1 - 1, \dots, 2\lambda_d - 2d + 1)$.

In this example, $\lambda = (5, 5, 3, 2, 2)$, $d = 3$ and $\varphi(\lambda) = (9, 7, 1)$.

$$\lambda \qquad\qquad\qquad \varphi(\lambda)$$

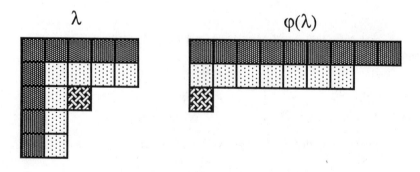

The definition of φ^{-1} is clear. Just "bend" each odd part in $\lambda \in POD(n)$ in the middle and center it on the main diagonal. Since the parts are distinct, this will form a Ferrers diagram of some partition in $PSC(n)$.

Is there a generating function identity implied by Theorem 3.3? Clearly,

$$(3.5) \qquad \sum_{n=0}^{\infty} |POD(n)| \, x^n \;=\; \prod_{i=0}^{\infty} (1 + x^{2i+1}).$$

For $|PSC(n)|$ we can use the decomposition by the $d \times d$ Durfee square to obtain

$$(3.6) \qquad \sum_{d=0}^{\infty} \frac{x^{d^2}}{(1-x^2) \cdots (1-x^{2d})} \;=\; \sum_{n=0}^{\infty} |PSC(n)| \, x^n.$$

since the Durfee square contains d^2 boxes, and what remains are two copies of some partition with at most d parts. So Theorem 3.3 is interpreted by

$$(3.7) \qquad \sum_{d=0}^{\infty} \frac{x^{d^2}}{(1-x^2) \cdots (1-x^{2d})} \;=\; \prod_{i=0}^{\infty} (1 + x^{2i+1}).$$

In fact, (3.7) is a special case of the analogue of the binomial theorem for q-binomial coefficients [An].

We shall return to partitions in Chapter 4.

§3.4 Permutations

Permutations are such naturally occurring objects that it should come as no surprise many interesting bijections involve them. We have already seen several properties of the inversion number of a permutation from the inversion poset in §2.1. We state two of these properties as bijections in this section. We also give some bijections related to Stirling numbers. Finally, we consider multiset permutations, which will be discussed in more detail in §§3.5-3.7.

In §1.1 we saw that the inversion sequence $(a_1, \dots, a_n)$ of a permutation π uniquely defined π. So the map φ from permutations of n to sequences $(a_1, \dots, a_n)$ with $0 \le a_i < i$ is a bijection. If $\varphi(\pi) = (a_1, \dots, a_n)$, clearly the number of inversions of π, $inv(\pi)$, is $a_1 + \dots + a_n$. Thus, the bijection φ proves the following theorem.

THEOREM 4.1 *The generating function for the number of inversions of permutations* π *of* n *is*

$$\sum_{\pi \in S_n} q^{inv(\pi)} = (1 + q)(1 + q + q^2) \cdots (1 + q + \ldots + q^{n-1}).$$

If we consider π as an element of the inversion poset $\mathcal{L}_n$, rank$(\pi) = inv(\pi)$, so that Theorem 4.1 also gives the generating function for the Whitney numbers (see Exercise 6 of Chapter 2) of $\mathcal{L}_n$. The rank symmetry and rank unimodality of $\mathcal{L}_n$ can be shown from this generating function (see Exercise 7 of Chapter 2).

Another interesting fact about $\mathcal{L}_n$ is that the inverse map preserves the partial order of $\mathcal{L}_n$. This means that if $\pi < \sigma$, then $\pi^{-1} < \sigma^{-1}$. This implies that the inverse map preserves the rank, so inv$(\pi) = inv(\pi^{-1})$. We prove this, based upon a bijection between permutations and non-attacking rooks on a chessboard.

THEOREM 4.2 *If* π *is a permutation of* n, *then* inv$(\pi) = inv(\pi^{-1})$.

Proof There is a bijection between permutations π of n and arrangements of n non-attacking rooks on an $n \times n$ chessboard: place the rook of row i in column $\pi(i)$. For example, here is the arrangement for $\pi = 23154$.

The value of inv(π) is the number of pairs (i, j) with $i < j$ and $\pi(i) > \pi(j)$. This corresponds to a pair of rooks (R_1, R_2), with R_1 to the right and above R_2. That means there is a square directly to the left of R_1 which is also directly above R_2. If we shade all squares to the left of a rook, and also all squares above a rook, these pairs are those squares which are shaded in both directions.

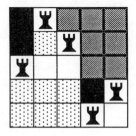

In this example, $inv(\pi) = 3$.

The arrangement of rooks corresponding to π^{-1} is precisely the <u>transpose</u> of the arrangement for π. Transposing the chessboard will not change the number of doubly shaded squares.

∎

In §1.1 we gave another representation of a permutation π: the decomposition of π into disjoint cycles. This gives us another bijection. We put the smallest number of each cycle at the end of that cycle, and put the cycles in order of their last entries. This defines the *canonical cycle decomposition* of π. The canonical cycle decomposition of $\pi = 463281795$ is $(4261)\ (3)\ (895)\ (7)$.

If we remove the parentheses from the canonical cycle decomposition of π, we have another permutation, $\phi(\pi) = 426138957$, in one line notation. The map ϕ is a bijection because we can recover π from $\phi(\pi)$: the first cycle of π is the initial segment of $\phi(\pi)$ ending at 1. The next cycle of π ends in the smallest number not appearing in the first. The remaining cycles are obtained in the same manner. For $n = 4$ the bijection is given below.

π	$\phi(\pi)$
1234	1234
1243	1243
1423	1432

4123	4321
4132	4213
1432	1423
1342	1342
1324	1324
3124	3214
3142	3421
3412	3142
4312	4231
4321	4132
3421	3241
3241	3412
3214	3124
2314	2314
2341	2341
2431	2413
4231	4123
4213	4312
2413	2431
2143	2143
2134	2134

The first entries of $\pi = (\pi_1, \ldots, \pi_n)$ and $\phi(\pi) = (\sigma_1, \ldots, \sigma_n) = \sigma$ are the same on this list. This is always the case because the last entry of the first cycle of π is 1, so π maps 1 to σ_1. This is $\pi_1 = \sigma_1$. We also notice that the *falls*, or *descents*, of σ must lie <u>inside</u> the cycles of π. The example $\sigma = 4 / 26 / 1389 / 57$ has 3 falls: 4 to 2, 6 to 1, and 9 to 5, which have been indicated with a slash. In general, i is a fall of σ if $\sigma_i > \sigma_{i+1}$. So, for any fall $\sigma_i \sigma_{i+1}$ of σ, π maps $j = \sigma_i$ to $m = \sigma_{i+1}$ ($\pi_j = m$), where $j > m$. Clearly any such j and m give a fall in σ. Thus we have the following theorem.

THEOREM 4.3 *The number of permutations π of n with k falls is equal to the number of permutations π of n whose two line notation has m below j with $j > m$ in exactly k positions.*

If a permutation σ has k falls, it also has $k + 1$ *runs*. A run of a permutation is the string of integers between two consecutive falls or between a fall and an end of the permutation. They are the strings separated by the slashes of the falls. Let $e(n, k)$, the *Eulerian number*, be the number of permutations σ of n with k runs. For $n = 4$, $e(4, 1) = 1$, $e(4, 2) = 11$, $e(4, 3) = 11$ and $e(4, 4) = 1$. These numbers have several properties which mimic those of the binomial coefficients or the Stirling numbers of the second kind.

The analogue of Pascal's triangle is

(4.1) $e(n, k) = k\, e(n - 1, k) + (n - k + 1)\, e(n - 1, k - 1).$

It is easy to give a bijective proof of (4.1). As in the proof of (2.3) in §1.2 or (5.1) in §1.5, just consider where n is placed in σ.

We can also classify permutations by the number of cycles. Let $c(n, k)$ denote the number of permutations of n with k cycles. For reasons which will become clear (see (4.3) and (4.5) below), this number is usually given a sign: $s(n, k) = (-1)^{n+k} c(n, k)$. For example, $s(4, 1) = -6$, $s(4, 2) = 7$, $s(4, 3) = -6$, and $s(4, 4) = 1$. The numbers $s(n, k)$ are called *Stirling numbers of the first kind*. They have many properties analogous to those of Stirling numbers of the second kind. For instance, they satisfy a three-term recurrence:

(4.2) $s(n, k) = -(n - 1)\, s(n - 1, k) + s(n - 1, k - 1).$

This recurrence can be proved in the same manner as (2.3) or (5.1) in Chapter 1 or (4.1) above. You are asked to prove (4.1) and (4.2) in Exercise 23.

Stirling numbers are related to one another by an *orthogonality formula*:

(4.3) $\displaystyle\sum_{k=0}^{n} S(n, k)\, s(k, j) = \begin{cases} 1 & \text{if } n = j \\ 0 & \text{if } n \neq j. \end{cases}$

You are asked to give a combinatorial proof of (4.3) in Chapter 4.

Stirling numbers and Eulerian numbers satisfy analogues of the binomial theorem. For Eulerian numbers, this is

(4.4) $\displaystyle x^n = \sum_{k=0}^{n} \binom{x+k-1}{n}\, e(n, k).$

For Stirling numbers of the first kind, it is

$$(4.5) \qquad (x)_n \;=\; \sum_{k=0}^{n} s(n, k)\, x^k.$$

For Stirling numbers of the second kind, it is

$$(4.6) \qquad x^n \;=\; \sum_{k=0}^{n} S(n, k)\, (x)_k.$$

(Recall from §1.1 that $(x)_n = x\,(x-1)\cdots(x-n+1)$.)

Like the binomial theorem, equations (4.4)-(4.6) have many proofs. We will concentrate on combinatorial proofs in the style of the two proofs of the binomial theorem given at the start of this chapter. The reader might want to review those two proofs at this time. Recall that in the first proof, we defined two *weighted sets*, $\mathcal{A}$ and $\mathcal{B}$, where the *weight* of an element was a monomial. Then the two sides of the equation corresponded to summing the weights of the elements of the two sets $\mathcal{A}$ and $\mathcal{B}$. We then constructed a *weight-preserving bijection* between the two sets.

In the second proof, we let x be a positive integer. Then we described a set which was counted in two different ways by the two sides of the identity.

Equations (4.4)-(4.6) cause some difficulty because they involve signs. We will learn other techniques for dealing with signs in Chapter 4. With weight-preserving bijections, the signs can be incorporated in the weights. Let's prove (4.5) in this way. First, we expand the left-hand side of (4.5):

$$(4.7) \qquad (x)_n \;=\; \sum_{A \subset [n]} \; \prod_{i \in A} (1-i)\, x^{n-|A|}.$$

From (4.7) we see that an appropriate set $\mathcal{A}$ would be all pairs (A, f) where $A \subset [n]$ and f is a function $f : A \to [n]$ such that $f(i) < i$, $i \in A$. The appropriate weight would be $w(A, f) = (-1)^{|A|} x^{n-|A|}$. The other set, $\mathcal{B}$, will be all permutations of n. The right-hand side of (4.5) tells us the weight of a permutation, $w(\pi) = (-1)^{n+k} x^k$, where k is the number of cycles in π.

We now need a weight-preserving bijection $\varphi : \mathcal{B} \to \mathcal{A}$. For $\pi \in \mathcal{B}$, the canonical cycle decomposition of π has k special entries, those at the end of each cycle. In $\varphi(\pi) = (A, f)$, let A be those entries of π which are not special. The function f indicates how the remaining entries of π were positioned. Define $f(i) - 1$ to be the number of entries to the left of i which are $< i$. For example, if $n = 9$, $A = \{2, 4, 6, 8, 9\}$, and $\pi = (62481)\,(93)\,(5)\,(7)$, then $f(2) = 1$, $f(4) = 2$, $f(6) = 1$, $f(8) = 4$, and $f(9) = 6$. It is not hard to prove that φ is a weight-preserving bijection.

Proofs of (4.4) and (4.6) are sketched in Exercises 27 and 26.

Now let's use the second method to prove (4.4)-(4.6). Remember that x is a positive integer. First consider (4.6). The left-hand side counts the set of all functions $f : [n] \to [x]$. Classify these functions by the partition of [n] given by f^{-1}. There are $S(n, k)$ possible partitions. How many functions from [n] to [x] have a given pre-image partition π? Clearly, elements of the same block map to the same element of [x], while elements of different blocks must be mapped to different elements of [x]. If π has k blocks, the number of such functions is just $(x)_k$.

Unfortunately, the terms in the sum in (4.5) alternate in sign. In Chapter 4 we shall learn more about dealing with signs. For now, we can replace x with $-x$ in (4.5) and multiply by $(-1)^n$ to get

$$(4.8) \qquad x(x+1)\cdots(x+n-1) = \sum_{k=0}^{n} c(n, k)\, x^k.$$

The left-hand side of (4.8) counts placements of n labeled balls in x labeled boxes, where the balls in any single box are ordered. For example, here is such a placement when $n = 9$ and $x = 4$.

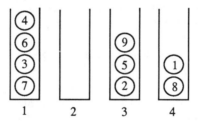

The right-hand side counts the same set in the following way. Construct a permutation of the balls and write it in cycle notation. Let k be the number of cycles. Now assign each cycle to a box. The number of ways of doing this is $c(n, k)\, x^k$, for each k. Within any single box there will be a collection of cycles. These form an ordering of the balls in that box. In the example, box 1 contains the permutation 7364, box 2 is empty, box 3 contains the permutation 259, and box 4 the permutation 81. So the resulting permutation is (81) (2) (743) (5) (6) (9), with (743) and (6) assigned to box 1; (2), (5) and (9) assigned to box 3; and (81) assigned to box 4.

We leave (4.4) as an exercise (Exercise 21).

The final topic of this section is multiset permutations. A *multiset* $\mathfrak{M}$ is a "set" of objects where repetitions are allowed. Usually the word "set" refers to distinct objects, so we use the word multiset. An example of a multiset is

$$\mathfrak{M} = \{A, A, N, N, O, R, R, S, S, T, U, U, Y\}.$$

The *multiplicity* of A in $\mathfrak{M}$ is 2, because $\mathfrak{M}$ contains 2 A's. A *multiset permutation* is some ordering of the elements of a multiset. TYRANNOSAURUS is a multiset permutation of $\mathfrak{M}$.

We often denote a multiset $\mathfrak{M}$ by the objects $\{1, \ldots, n\}$ with multiplicities $(m_1, \ldots, m_n)$. It is clear that the number of multiset permutations of $\mathfrak{M}$ is $M! / m_1! \cdots m_n!$, where $M = m_1 + \ldots + m_n$. This number is called the *multinomial coefficient*

$$\binom{M}{m_1, \ldots, m_n}$$

because of the multinomial theorem

$$(x_1 + \ldots + x_n)^M = \sum_{(m_1, \ldots, m_n)} \binom{M}{m_1, \ldots, m_n} x_1^{m_1} \cdots x_n^{m_n}$$

Multiset permutations are naturally given in one-line notation. There is also a two-line notation. For $\pi = 221231122311$ we could write

$$\pi = \begin{pmatrix} 1 & 2 & 3 & 4 & 5 & 6 & 7 & 8 & 9 & 10 & 11 & 12 \\ 2 & 2 & 1 & 2 & 3 & 1 & 1 & 2 & 2 & 3 & 1 & 1 \end{pmatrix}$$

to signify that 2 is first, 3 is fifth, and so on. Another two-line notation places the same entries on the top line in increasing order,

$$\pi = \begin{pmatrix} 1 & 1 & 1 & 1 & 1 & 2 & 2 & 2 & 2 & 2 & 3 & 3 \\ 2 & 2 & 1 & 2 & 3 & 1 & 1 & 2 & 2 & 3 & 1 & 1 \end{pmatrix}$$

so that both two-line notations give the usual two-line notation if the multiset permutation is a permutation of n. An analogue of the cycle representation of π also exists (see [Lot], Chapter 10).

The number of inversions of a multiset permutation, $\text{inv}(\pi)$, is the number of (i, j) with $i < j$ and $\pi_i > \pi_j$. In our example, $\text{inv}(\pi) = 26$. You are asked to investigate this statistic in Exercise 29.

§3.5 Tableaux

The tableau is a fundamental construction in the classical presentation of representations of the symmetric group. But its algebraic importance carries over to many other areas of mathematics: symmetric functions, invariant theory, algebraic geometry, Lie algebras and combinatorics, to name a few. Two practical applications are in quantum theory (representations of $GL(n)$) and chemistry (Polya counting theory). In this section we shall give a basic bijection for column strict tableaux, and relate these tableaux to Young's lattice of §2.1. They will play a central role in the next two sections on the Schensted correspondence.

Let λ be a partition of n. A *tableau* T of *shape* λ is the Ferrers diagram of λ with each cell filled with a positive integer. These positive integers are called the *entries* of the tableau T. The *content* $\rho = (\rho_1, \dots, \rho_m)$ of a tableau T is the vector of multiplicities of the entries of the tableau T. This means that T has ρ_1 1's, ρ_2 2's, $\dots$, and ρ_m m's. When it is necessary, we will append zeros to the end of ρ.

The tableau T below has shape $\lambda = 4\,2^2$ and content $\rho = (2, 0, 3, 1, 2)$.

$$T = \begin{array}{|c|c|c|c|} \hline 3 & 3 & 5 & 1 \\ \hline 5 & 1 \\ \cline{1-2} 4 & 3 \\ \cline{1-2} \end{array}$$

We will be concerned with a special kind of tableau. A tableau T is called *column strict* if the entries of T are non-decreasing along the rows of T and strictly increasing down the columns of T. The tableau below of shape $\lambda = 4\,2^2$ and content $\rho = (2, 0, 3, 1, 2)$ is column strict.

$$T = \begin{array}{|c|c|c|c|} \hline 1 & 1 & 3 & 5 \\ \hline 3 & 3 \\ \cline{1-2} 4 & 5 \\ \cline{1-2} \end{array}$$

Let $\mathcal{J}(\lambda, \rho)$ be the set of all column strict tableaux of shape λ and content ρ. For example, $\mathcal{J}(3\,2, (2, 1, 1, 1))$ consists of the following tableaux

$$\begin{array}{|c|c|c|} \hline 1 & 1 & 2 \\ \hline 3 & 4 \\ \cline{1-2} \end{array} \qquad \begin{array}{|c|c|c|} \hline 1 & 1 & 3 \\ \hline 2 & 4 \\ \cline{1-2} \end{array} \qquad \begin{array}{|c|c|c|} \hline 1 & 1 & 4 \\ \hline 2 & 3 \\ \cline{1-2} \end{array}$$

while $\mathfrak{J}(3\,2, (1, 2, 1, 1))$ has

$$\begin{array}{|c|c|c|}\hline 1 & 2 & 2 \\ \hline 3 & 4 & \\ \cline{1-2}\end{array} \qquad \begin{array}{|c|c|c|}\hline 1 & 2 & 3 \\ \hline 2 & 4 & \\ \cline{1-2}\end{array} \qquad \begin{array}{|c|c|c|}\hline 1 & 2 & 4 \\ \hline 2 & 3 & \\ \cline{1-2}\end{array} \quad .$$

It is not an accident that the number of tableaux in each set is the same. It is our next theorem.

THEOREM 5.1 *There is bijection between* $\mathfrak{J}(\lambda, \rho)$ *and* $\mathfrak{J}(\lambda, \rho')$, *where* ρ' *is obtained from* ρ *by applying an adjacent transposition.*

Proof Let ρ' agree with ρ, except for ρ_k and ρ_{k+1}, which have been interchanged. Let $T \in \mathfrak{J}(\lambda, \rho)$ so that T has ρ_k k's and ρ_{k+1} $k+1$'s. We need to produce a tableau $T' \in \mathfrak{J}(\lambda, \rho')$ which has ρ_k $k+1$'s and ρ_{k+1} k's. We will do this by switching some k's in T to $k+1$'s, and also switching some $k+1$'s in T to k's. The k and $k+1$ entries in row i of T have the following structure.

$$\begin{array}{lcccc}
\text{row } i-1 & & & & k \cdots k \\
\text{row } i & \overbrace{k \cdots k}^{a} & \overbrace{k \cdots k}^{b} & \overbrace{k+1 \cdots k+1}^{c} & \underbrace{k+1 \cdots k+1}_{d} \\
\text{row } i+1 & k+1 \cdots k+1 & & &
\end{array}$$

For this row, we let $a \geq 0$ be the number of k's with $k+1$'s below them. Immediately to the right of these k's, there will be some number $b \geq 0$ of k's with no $k+1$'s below, then some number $c \geq 0$ of $k+1$'s with no k above, and finally some number $d \geq 0$ of $k+1$'s with k's above them.

We change T to T' by changing each row i to this form.

$$\begin{array}{lcccc}
\text{row } i-1 & & & & k \cdots k \\
\text{row } i & \overbrace{k \cdots k}^{a} & \overbrace{k \cdots k}^{c} & \overbrace{k+1 \cdots k+1}^{b} & \underbrace{k+1 \cdots k+1}_{d} \\
\text{row } i+1 & k+1 \cdots k+1 & & &
\end{array}$$

Note that the k's and $k+1$'s which are paired with $k+1$'s below and k's above are left unchanged. However, the b k's have become c k's, and the c $k+1$'s have become b $k+1$'s. So the total number of unpaired k's in T is equal to the total number of unpaired $k+1$'s in T'. This implies that T' has ρ_{k+1} k's and ρ_k $k+1$'s. It is easy to see that T' is column strict.

It is clear that if we apply this map again we obtain T, that is T" = T. So the map T → T' is a bijection.

An example of this bijection is $T \in \mathfrak{I}(8^2\,4\,2, (3, 6, 4, 1, 2, 6))$,

$$T \qquad
\begin{array}{|c|c|c|c|c|c|c|c|}
\hline
1 & 1 & 1 & 2 & 2 & 2 & 2 & 3 \\
\hline
2 & 2 & 3 & 3 & 4 & 6 & 6 & 6 \\
\hline
\multicolumn{1}{|c|}{3} & \multicolumn{1}{c|}{5} & \multicolumn{1}{c|}{6} & \multicolumn{1}{c|}{6} \\
\cline{1-4}
\multicolumn{1}{|c|}{5} & \multicolumn{1}{c|}{6} \\
\cline{1-2}
\end{array}$$

and $T' \in \mathfrak{I}(8^2\,4\,2, (3, 4, 6, 1, 2, 6))$,

$$T' \qquad
\begin{array}{|c|c|c|c|c|c|c|c|}
\hline
1 & 1 & 1 & 2 & 2 & 3 & 3 & 3 \\
\hline
2 & 2 & 3 & 3 & 4 & 6 & 6 & 6 \\
\hline
\multicolumn{1}{|c|}{3} & \multicolumn{1}{c|}{5} & \multicolumn{1}{c|}{6} & \multicolumn{1}{c|}{6} \\
\cline{1-4}
\multicolumn{1}{|c|}{5} & \multicolumn{1}{c|}{6} \\
\cline{1-2}
\end{array} \qquad .$$

The paired 2's and 3's have been boxed in bold face.

Because of Theorem 1.3 of Chapter 1, we immediately have this corollary.

COROLLARY 5.2 *If ρ' is any reordering of ρ, then $|\mathfrak{I}(\lambda, \rho)| = |\mathfrak{I}(\lambda, \rho')|$.*

From Corollary 5.2, we can reorder ρ so that $\rho_i \geq \rho_{i+1}$. This means that we can assume that ρ is a partition. The number of column strict tableaux of shape λ and content ρ, $|\mathfrak{I}(\lambda, \rho)|$, is called the Kostka number $K_{\lambda\rho}$. It plays a key role in the study of tableaux and their connections to various branches of mathematics.

A column strict tableau of content $\rho = (1, 1, \ldots , 1)$ is called a standard tableau. There are 5 standard tableaux of shape 3 2.

$$
\begin{array}{|c|c|c|}
\hline
1 & 2 & 3 \\
\hline
4 & 5 \\
\cline{1-2}
\end{array} \quad
\begin{array}{|c|c|c|}
\hline
1 & 2 & 4 \\
\hline
3 & 5 \\
\cline{1-2}
\end{array} \quad
\begin{array}{|c|c|c|}
\hline
1 & 2 & 5 \\
\hline
3 & 4 \\
\cline{1-2}
\end{array} \quad
\begin{array}{|c|c|c|}
\hline
1 & 3 & 4 \\
\hline
2 & 5 \\
\cline{1-2}
\end{array} \quad
\begin{array}{|c|c|c|}
\hline
1 & 3 & 5 \\
\hline
2 & 4 \\
\cline{1-2}
\end{array}
$$

The number of standard tableaux of shape λ, $K_{\lambda\,1^n}$, is also very important and is denoted d_λ. The next theorem is clear.

THEOREM 5.3 *The number of standard tableaux of shape* λ, d_λ, *equals the number of maximal chains in Young's lattice* $\mathcal{Y}_\lambda$.

There is an amazing formula for d_λ. It is simple to state and difficult to prove. We will give a proof of an equivalent formula in §4.5. To state this formula, we need to define a hook. Let c be a cell of the Ferrers diagram of λ. We write $c \in \lambda$. The *hook* of c, H_c, consists of the cells to the right of c, below c, and c itself. In the bijection proving Theorem 3.3 of this chapter, we used the hooks of the major diagonal. The length of the hook of c, h_c, is $|H_c|$. In the example below the hook of c is shaded and its length $h_c = 7$.

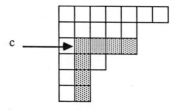

The formula for d_λ is called the *hook formula*.

THEOREM 5.4 *Let* λ *be a partition of* n. *Then*

$$d_\lambda = \frac{n!}{\prod_{c \in \lambda} h_c} .$$

As an example, take $\lambda = 4\, 2^2$. We insert the hook lengths into the cells of λ.

6	5	2	1
3	2		
2	1		

Then

$$d_{422} = \frac{8!}{6 \cdot 5 \cdot 2 \cdot 1 \cdot 3 \cdot 2 \cdot 2 \cdot 1} = 56.$$

§3.6 The Schensted Correspondence

The construction which plays a central role in the applications of tableaux to various areas of mathematics is the *Schensted correspondence*. This correspondence is a bijection between multiset permutations and pairs of tableaux of the same shape: one standard, the other column strict. We shall find that several properties of permutations manifest themselves in the tableaux via the Schensted correspondence.

First we consider a special case: permutations instead of multiset permutations. Then the Schensted correspondence becomes this theorem.

THEOREM 6.1 *The Schensted correspondence is a bijection between all permutations* π *of* n *and all pairs* (P, Q) *of standard tableaux of the same shape* λ. *The shape* λ *is an arbitrary partition of* n.

For $n = 3$ the bijection is

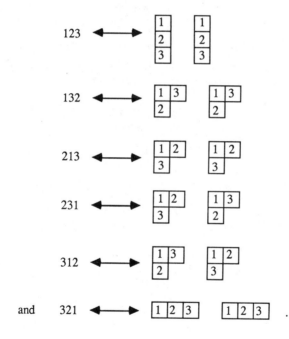

The Schensted correspondence is built inductively. First we take the example $\pi = 35186724$ as a permutation of 8. Clearly, if we delete 4 from π, we have a string π' of length 7. If we subtract one from all of the entries of π' which are > 4, we have the permutation $\pi'' = 3417562$ of 7. By induction, π'' corresponds to a pair of standard tableaux (P'', Q'') with 7 cells. What we need to do is this:

insert the number 4 into P'' and Q'' (in some way) and add one to all of the entries of P'' and Q'' which are ≥ 4. This will give us P and Q. Of course, the difficult part of this is to determine exactly how to insert the number 4. This algorithm is called the *Schensted column insertion* algorithm.

First, we should remark that it is not necessary to do the addition and subtraction of one to the entries of P'' and Q''. The Schensted correspondence will produce tableaux (P', Q') for π', where the entries of P' are [8] − {4}, and the entries of Q' are [7]. For π = 35186724,

$$\pi' = 3518672, \quad P' = \begin{array}{|c|c|c|} \hline 1 & 3 & 8 \\ \hline 2 & 5 \\ \cline{1-2} 6 \\ \cline{1-1} 7 \\ \cline{1-1} \end{array} \qquad Q' = \begin{array}{|c|c|c|} \hline 1 & 3 & 7 \\ \hline 2 & 5 \\ \cline{1-2} 4 \\ \cline{1-1} 6 \\ \cline{1-1} \end{array} \;.$$

The Schensted column insertion algorithm inserts 4 into P' and 8 into Q'. The natural position of 4 in the first column of P' is between the 2 and the 6. The 4 takes the place of the 6, or *bumps* the 6 out of the first column of P'. The 6 now is inserted by the same method into the second column of P'. This time the 6 is placed after the 5 and does not bump any entry of the second column. The column insertion algorithm has been completed. The resulting tableau P is

$$P = \begin{array}{|c|c|c|} \hline 1 & 3 & 8 \\ \hline 2 & 5 \\ \cline{1-2} 4 & 6 \\ \cline{1-2} 7 \\ \cline{1-1} \end{array} \;.$$

Note that the shape of P differs from that of P' by the addition of exactly one cell (the cell 6 of P). We place an 8 in that cell for the definition of Q'.

$$Q = \begin{array}{|c|c|c|} \hline 1 & 3 & 7 \\ \hline 2 & 5 \\ \cline{1-2} 4 & 8 \\ \cline{1-2} 6 \\ \cline{1-1} \end{array}$$

A given insertion could cause several bumps. An example in the general case is given later.

To find P and Q from π, just apply the column insertion algorithm successively to the entries of π. If we use two line notation for π,

$$\pi \;=\; \begin{pmatrix} 1\,2\,3\,4\,5\,6\,7\,8 \\ 3\,5\,1\,8\,6\,7\,2\,4 \end{pmatrix} \xleftarrow{\hspace{1cm}} \text{entries of Q}$$
$$\xleftarrow{\hspace{1cm}} \text{entries of P}$$

the entries of the top line are successively inserted into Q, while those of the bottom line are likewise inserted into P. What follows is the sequence of tableaux obtained from the column insertions of π to arrive at P' and Q'.

3	3	1

35

$$\begin{array}{|c|} \hline 3 \\ \hline 5 \\ \hline \end{array} \qquad \begin{array}{|c|} \hline 1 \\ \hline 2 \\ \hline \end{array}$$

351

$$\begin{array}{|c|c|} \hline 1 & 3 \\ \hline 5 \\ \cline{1-1} \end{array} \qquad \begin{array}{|c|c|} \hline 1 & 3 \\ \hline 2 \\ \cline{1-1} \end{array}$$

3518

$$\begin{array}{|c|c|} \hline 1 & 3 \\ \hline 5 \\ \cline{1-1} 8 \\ \cline{1-1} \end{array} \qquad \begin{array}{|c|c|} \hline 1 & 3 \\ \hline 2 \\ \cline{1-1} 4 \\ \cline{1-1} \end{array}$$

35186

$$\begin{array}{|c|c|} \hline 1 & 3 \\ \hline 5 & 8 \\ \hline 6 \\ \cline{1-1} \end{array} \qquad \begin{array}{|c|c|} \hline 1 & 3 \\ \hline 2 & 5 \\ \hline 4 \\ \cline{1-1} \end{array}$$

351864

$$\begin{array}{|c|c|c|} \hline 1 & 3 & 8 \\ \hline 4 & 5 \\ \cline{1-2} 6 \\ \cline{1-1} \end{array} \qquad \begin{array}{|c|c|c|} \hline 1 & 3 & 6 \\ \hline 2 & 5 \\ \cline{1-2} 4 \\ \cline{1-1} \end{array}$$

3518647

$$\begin{array}{|c|c|c|} \hline 1 & 3 & 8 \\ \hline 4 & 5 \\ \cline{1-2} 6 \\ \cline{1-1} 7 \\ \cline{1-1} \end{array} \qquad \begin{array}{|c|c|c|} \hline 1 & 3 & 6 \\ \hline 2 & 5 \\ \cline{1-2} 4 \\ \cline{1-1} 7 \\ \cline{1-1} \end{array}$$

If the Schensted correspondence is a bijection, we must be able to recover π from (P, Q). It is clear what we do. Find the largest cell c in Q (in the example it contains 8), and find the entry e of that cell in P (in the example it is 6). Now we insert e in the left neighboring column, and bump the largest entry of that column which is $<$ e to the left. (This is exactly the reverse of the bumping procedure.) We continue this procedure until an entry f is bumped out of the first column of P, creating the new tableau P'. Then f is the last entry of π (in the example $f = 4$).

Removing c from Q gives Q'. Next, we start with the largest entry of Q', and do this inverse bumping procedure until another entry is bumped from the first column of P', and continue.

We will not give a formal proof by induction of this case. Instead, we will consider the *generalized Schensted correspondence* for multiset permutations. Let $\mathfrak{M}$ be a multiset with *content* $(\rho_1, \ldots, \rho_m)$, so that $\mathfrak{M}$ contains ρ_i i's. A permutation π of $\mathfrak{M}$ is any sequence of the elements of $\mathfrak{M}$. For $(\rho_1, \ldots, \rho_m) = (1, 4, 2)$, one such π is 2322132. Clearly if each $\rho_i = 1$, $\mathfrak{M}$ has no repetitions so that multiset permutations are just usual permutations. The generalized Schensted correspondence will associate to π the pair of tableaux (P, Q), where P is column strict of content ρ, and Q is a standard tableau of the same shape as P. For $\pi = 2322132$ we will see that

$$(P, Q) = \left(\begin{array}{|c|c|c|c|} \hline 1 & 2 & 2 & 2 \\ \hline 2 & 3 & 3 \\ \cline{1-3} \end{array} \ , \ \begin{array}{|c|c|c|c|} \hline 1 & 3 & 4 & 5 \\ \hline 2 & 6 & 7 \\ \cline{1-3} \end{array} \right).$$

To produce P and Q, we use the same bumping procedure with a minor modification. Suppose we are inserting k into a column. The number which is bumped out of the column is the smallest number $\geq k$. (In the previous case, it was $> k$ because repeats were not allowed.) For example, suppose we are inserting 4 into the column strict tableau P.

$$P = \begin{array}{|c|c|c|c|c|c|c|c|} \hline 1 & 1 & 1 & 2 & 4 & 4 & 6 & 6 \\ \hline 2 & 3 & 3 & 3 & 5 & 7 \\ \cline{1-6} 3 & 5 & 5 & 6 & 7 \\ \cline{1-5} 4 & 7 & 7 & 7 \\ \cline{1-4} 6 \\ \cline{1-1} \end{array}$$

Then the 4 will bump the 4 in the first column, so that the new first column is the old first column.

$$\begin{array}{|c|} \hline 1 \\ \hline 2 \\ \hline 3 \\ \hline 4 \\ \hline 6 \\ \hline \end{array}$$

The bumped 4 now bumps the 5 in the second column, for the following first two new columns.

1	1
2	3
3	4
4	7
6	

The 5 from column 2 bumps the 5 in column 3.

1	1	1
2	3	3
3	4	5
4	7	7
6		

The 5 from column 3 bumps the 6 from column 4.

1	1	1	2
2	3	3	3
3	4	5	5
4	7	7	7
6			

The 6 from column 4 bumps the 7 from column 5.

1	1	1	2	4
2	3	3	3	5
3	4	5	5	6
4	7	7	7	
6				

The 7 from column 5 bumps the 7 from column 6.

1	1	1	2	4	4
2	3	3	3	5	7
3	4	5	5	6	
4	7	7	7		
6					

The 7 from column 6 is larger than everything in column 7. It is therefore placed at the end of column 7, and the insertion of 4 into P has been completed.

1	1	1	2	4	4	6	6
2	3	3	3	5	7	**7**	
3	4	5	5	6			
4	7	7	7				
6							

The new cell has been marked in bold face. We would place the next entry of Q (here a 25) in that cell.

The reader should verify that $\pi = 2322132$ gives the (P, Q) that was previously claimed. The entries of Q are 1234567, the first row of the two line notation for π.

The inverse Schensted correspondence $(P, Q) \to \pi$ is as before. We use the largest entry of Q to find the entry e of P which bumps to the left. This time the largest number of the column which is $\le e$ is bumped to the left. (Note that the entry directly to the left of e is $\le e$, so that this set is non-empty.) The number eventually bumped out of the first column is again the last entry of π. We call this inverse procedure *column deletion*. Because column deletion is the inverse to column insertion, it is easy to see by induction that the generalized Schensted correspondence is a bijection.

THEOREM 6.2 *The generalized Schensted correspondence is a bijection between all multiset permutations π of content ρ, and pairs of tableaux (P, Q), where P is column strict of content ρ, and Q is standard with the same shape as P.*

Suppose that $\rho = (\rho_1, \dots, \rho_m)$, so that the multinomial coefficient gives the number of multiset permutations π. The number of ordered pairs (P, Q) is a sum of Kostka numbers $K_{\lambda\rho}$ times d_λ, so that we have this corollary.

COROLLARY 6.3 *If $\rho = (\rho_1, \dots, \rho_m)$ and $\rho_1 + \dots + \rho_m = n$, then*

$$\binom{\rho_1 + \dots + \rho_m}{\rho_1, \dots, \rho_m} = \sum_\lambda K_{\lambda\rho} d_\lambda ,$$

where the summation is over all partitions λ of n.

For permutations of n, Corollary 6.3 specializes to the following corollary.

COROLLARY 6.4 *For any positive integer* n,

$$n! = \sum_{\lambda} d_{\lambda}^2$$

where the summation is over all partitions λ *of* n.

We now give the algorithms for Schensted column insertion, Schensted column deletion, the generalized Schensted correspondence (called Schensted encode) and the inverse generalized Schensted correspondence (called Schensted decode). For column insertion, the tableau is P with cell entries $P(i, j)$. The value to be inserted is k. The length of the jth column of P is c_j. The tableau with k inserted into P is $Ins(k, P)$; the new cell is $Cell(k, P)$.

ALGORITHM 17: *Schensted Column Insertion*

begin
    $P' \leftarrow P$
    DoMore $\leftarrow$ **true**
    $j \leftarrow 1$
    while DoMore **do**
        if $k \leq P'(c_j, j)$ **then**
            $i \leftarrow c_j$
            repeat
                $i \leftarrow i - 1$
            until $P'(i, j) < k$
            $i \leftarrow i + 1$
            $x \leftarrow P'(i, j)$
            $P'(i, j) \leftarrow k$
            $k \leftarrow x$
            $j \leftarrow j + 1$
        else
            $i \leftarrow c_j + 1$
            $P'(i, j) \leftarrow k$
            DoMore $\leftarrow$ **false**
    $Ins(P, k) \leftarrow P'$

$$Cell(P, k) \leftarrow (i, j)$$

end.

The column deletion algorithm begins at row r and column c of the tableau P. The tableau after deletion is Del((r, c), P) and the value bumped out is Val((r, c), P).

ALGORITHM 18: *Schensted Column Deletion*

begin

$$P' \leftarrow P$$
$$x \leftarrow P(r, c)$$
for $j \leftarrow c - 1$ **downto** 1 **do**
$$\quad i \leftarrow 1$$
$$\quad \textbf{repeat}$$
$$\quad\quad i \leftarrow i + 1$$
$$\quad \textbf{until } P'(i, j) > x \textbf{ or } i > c_j$$
$$\quad i \leftarrow i - 1$$
$$\quad y \leftarrow P'(i, j)$$
$$\quad P'(i, j) \leftarrow x$$
$$\quad x \leftarrow y$$
$$Del((r, c), P) \leftarrow P'$$
$$Val((r, c), P) \leftarrow y$$

end.

Now we give the Schensted encode algorithm. We use Ins(k, P) and Cell(k, P) from Algorithm 17. The permutation is π, whose two-line notation has top row $j_1, j_2, \ldots, j_n$ and bottom row $\pi_1, \pi_2, \ldots, \pi_n$. The resulting pair of tableaux is $(Sch_P(\pi), Sch_Q(\pi))$.

ALGORITHM 19: *Schensted Encode*

begin

$$P \leftarrow \varnothing$$
$$Q \leftarrow \varnothing$$
for $i \leftarrow 1$ **to** n **do**
$$\quad P \leftarrow Ins(\pi_i, P)$$
$$\quad Q(Cell(\pi_i, P)) \leftarrow j_i$$
$$Sch_P(\pi) \leftarrow P$$

$$\text{Sch}_Q(\pi) \leftarrow Q$$

end.

For the Schensted decode algorithm, we use $\text{Del}((i, j), P)$ and $\text{Val}((i, j), P)$ from Algorithm 18. $\text{MaxCell}(Q)$ computes the row and column containing the largest entry in Q. From the pair of tableaux (P, Q), this algorithm computes the permutation in two-line notation: $\text{Sch}_\pi(P, Q)$ is the bottom row and $\text{Sch}_\sigma(P, Q)$ the top row.

ALGORITHM 20: *Schensted Decode*

begin

        $P' \leftarrow P$

        $Q' \leftarrow Q$

        for $i \leftarrow n$ **downto** 1 **do**

            $c \leftarrow \text{MaxCell}(Q')$

            $\sigma_i \leftarrow Q'(c)$

            $P' \leftarrow \text{Del}(c, P')$

            $\pi_i \leftarrow \text{Val}(c, P')$

            $Q' \leftarrow Q' - \{c\}$

        $\text{Sch}_\pi(P, Q) \leftarrow \pi$

        $\text{Sch}_\sigma(P, Q) \leftarrow \sigma$

end.

Sometimes we will use the following notation and terminology. We call P the *P-tableau*, and Q the *Q-tableau*, if π corresponds to (P, Q). We also use the notation $P = \text{Sch}_P(\pi)$ and $Q = \text{Sch}_Q(\pi)$, as in Algorithm 19 above.

Several more remarkable properties of the Schensted correspondence are given in the next section.

§3.7 Properties of the Schensted Correspondence

In this section we shall investigate three of the many remarkable properties of the Schensted correspondence. The first property motivated Schensted's original paper [Sch]: given a permutation π of n, how can one find the length of the longest increasing subsequence of π? While solving that problem, we will also answer this question: what permutation corresponds to (Q, P) if π corresponds to (P, Q)? Finally, we will see that the matching in the Boolean algebra of §2.2 can be derived from the generalized Schensted correspondence.

The answer to the first question is given by this theorem.

THEOREM 7.1 *The number of rows of* P *(or* Q) *is the length of the longest increasing subsequence of* π.

In the example π = 35186724 of §3.6, P had 4 rows, so π has an increasing subsequence of length 4, 3567, and none longer.

To prove Theorem 7.1, we need to describe what happens to P and Q <u>in the first column only</u> at each stage. Suppose, as in the previous section, π = 35186724.

Entry into Q	Action Taken	Column 1 of P	of Q
1	Insert 3; $Q_{11} \leftarrow 1$	3	1
2	Insert 5; $Q_{21} \leftarrow 2$	3	1
		5	2
3	Insert 1; bump 3	1	1
		5	2
4	Insert 8; $Q_{31} \leftarrow 4$	1	1
		5	2
		8	4
5	Insert 6; bump 8	1	1
		5	2
		6	4
6	Insert 7; $Q_{41} \leftarrow 6$	1	1
		5	2
		6	4
		7	6
7	Insert 2; bump 5	1	1
		2	2
		6	4
		7	6
8	Insert 4; bump 6	1	1
		2	2

$$\begin{matrix} 4 & \quad & 4 \\ 7 & \quad & 6 \end{matrix}$$

The remaining columns of P and Q can be found by applying the Schensted correspondence to $\pi = 3856$ (the bumped numbers), using 3578 as the available entries for Q.

We call the "permutation"

$$\pi b \; = \; \begin{pmatrix} 3\,5\,7\,8 \\ 3\,8\,5\,6 \end{pmatrix}$$

the *bumping permutation* of π. To find the second columns of P and Q we apply the column insertion algorithm to this two line array.

Entry into Q	Action taken	Column 2 of P	of Q
3	Insert 3; $Q_{12} \leftarrow 3$	3	3
5	Insert 8; $Q_{22} \leftarrow 5$	3	3
		8	5
7	Insert 5; bump 8	3	3
		5	5
8	Insert 6; $Q_{32} \leftarrow 8$	3	3
		5	5
		6	8

Clearly, for the third column, P has 8 and Q has 7.

We now divide the n elements of the set $\{(i, \pi_i) : 1 \le i \le n\}$ into classes. We say that (i, π_i) is in *class* t if π_i was inserted into row t of column 1. In the example

$$\text{class } 1 = \{(1, 3), (3, 1)\},$$
$$\text{class } 2 = \{(2, 5), (7, 2)\},$$
$$\text{class } 3 = \{(4, 8), (5, 6), (8, 4)\},$$
and $$\text{class } 4 = \{(6, 7)\}.$$

We see that if (i, π_i) is in class t, the first column of P must have $t - 1$ entries smaller than π_i when π_i is inserted. So π has an increasing subsequence of length t which ends at π_i. Any longer such subsequence would force π_i to be inserted below row t. This property characterizes the class of a pair.

LEMMA 7.2 *The pair* (i, π_i) *belongs to class* t *if and only if the length of the largest increasing subsequence of* π *ending at* π_i *is* t.

Proof It remains to show if the length of the largest increasing subsequence of π ending at π_i is t, then (i, π_i) belongs to class t. We do this by induction on t. If t $= 1$, then π_i is smaller than all of the preceding π_j, so π_i is inserted into the first row of column and (i, π_i) belongs to class 1.

Now suppose $t > 1$, and choose an increasing subsequence S of π ending at π_i of length t. Let π_j be the predecessor to π_i in S. Then the subsequence S' = $S - \{\pi_i\}$ which ends at π_j is also of maximal length. By induction, (j, π_j) belongs to class $t - 1$. So when π_i was inserted into the first column, the entry v in row $t-1$ was $\leq \pi_j < \pi_i$. Thus, π_i was inserted either below row t or in row t. Let w be the entry in row t when π_i was inserted. The entry w (which precedes π_i) is a member of a pair in class t. By the first part of the theorem, there is an increasing subsequence of π of length t ending at w. If $w < \pi_i$, we could attach π_i to this subsequence and have an increasing subsequence of length $t + 1$. This contradicts our hypothesis, so $w > \pi_i$ and π_i was inserted into row t.

Proof of Theorem 7.1 Suppose S is one of the longest increasing subsequences of π and has length t and suppose that P has r rows. By Lemma 7.2, the last member π_i of S is inserted into row t of P, so $t \leq r$. Conversely, the (r, 1) entry of P is in class r, so Lemma 7.2 implies that $r \leq t$.

The answer to our second question is provided by Theorem 7.3.

THEOREM 7.3 *Suppose that* π *corresponds to* (P, Q) *in the Schensted correspondence. Then* π^{-1} *(the inverse of* π) *corresponds to* (Q, P).

Proof In fact, Theorem 7.3 holds for all two line arrays with distinct entries. In this case π^{-1} is the two line array obtained by interchanging the two lines, then sorting the columns according to the first line. Our proof will be by induction on the number of columns of P. First we show that the first columns agree, and then apply induction. We initially consider the permutation case, but the reader should have no difficulty extending to the two line case.

Let (i, π_i) belong to class t for π. It is clear from Lemma 7.2 that (π_i, i) is a pair for π^{-1} which also belongs to class t.

Note that if (i, π_i) and (j, π_j) are in the same class t for π with $i < j$, then $\pi_i > \pi_j$. This is because the entries in row t, column 1 must be non-increasing as the entries of π are inserted into P. This also implies that the smallest π_i in class t is the entry P_{t1}. The largest π_i in class t occurs when that cell is first occupied, so $Q_{t1} = i$. (In the example, the smallest π_i of class 3 is $\pi_8 = 4$, and $P_{31} = 4$. The largest π_i is $\pi_4 = 8$, and $Q_{31} = 4$.) Thus the subsequence S of π of the class t π_i's is decreasing and has last entry P_{t1}. The index i of the first π_i in S is the entry in Q_{t1}. The diagram below shows S, the entries P_{t1} and Q_{t1}, and the reversal of S for π^{-1}. This subsequence of π^{-1} gives the entries P'_{t1} and Q'_{t1} if π^{-1} corresponds to (P', Q').

$$
Q_{t1} \longrightarrow \begin{pmatrix} \boxed{i_1} < i_2 \ \ldots \ < i_k \\ \pi_{i_1} > \pi_{i_2} \cdots > \boxed{\pi_{i_k}} \end{pmatrix} \longleftarrow P_{t1} \qquad \begin{array}{l} \text{class } t \\ \text{pairs for } \pi \end{array}
$$

$$
Q'_{t1} \longrightarrow \begin{pmatrix} \boxed{\pi_{i_k}} < \pi_{i_{k-1}} \cdots < \pi_{i_1} \\ i_k > i_{k-1} \ \cdots > \boxed{i_1} \end{pmatrix} \longleftarrow P'_{t1} \qquad \begin{array}{l} \text{class } t \\ \text{pairs for } \pi^{-1} \end{array}
$$

This diagram shows that $Q'_{t1} = P_{t1}$ and $P'_{t1} = Q_{t1}$, so that the first columns of Q' and P, and P' and Q, are identical.

Next, suppose that the bumping permutation of π is πb. The entries on the first line of πb are used for the remaining columns of Q. Thus, they will also be used for the remaining columns of P'. The same statement can be made about the second line of πb and the remaining columns of P and Q'. So, if we show that πb is the inverse of the two line array $\pi^{-1}b$, the induction hypothesis will show that the P tableau of πb is the Q tableau of $\pi^{-1}b$, and vice versa. This will complete the proof.

In the class t pairs of π, π_{i_2} bumps π_{i_1}, π_{i_3} bumps π_{i_2}, etc. So πb contains the pairs

$$
\begin{pmatrix} i_2 & < \ldots < & i_k \\ \pi_{i_1} & > \ldots > & \pi_{i_{k-1}} \end{pmatrix}
$$

while $\pi^{-1}b$ contains the pairs from class t

$$\left(\begin{matrix} \pi_{i_{k-1}} & < & \cdots & < & \pi_{i_1} \\ i_k & > & \cdots & > & i_2 \end{matrix} \right).$$

For example, the class 3 pairs of $\pi = 35186724$ give

$$\left(\begin{matrix} \boxed{4} & 5 & 8 \\ 8 & 6 & \boxed{4} \end{matrix} \right) \longrightarrow \left(\begin{matrix} 5 & 8 \\ 8 & 6 \end{matrix} \right),$$

while the class 3 pairs of π^{-1} give

$$\left(\begin{matrix} \boxed{4} & 6 & 8 \\ 8 & 5 & \boxed{4} \end{matrix} \right) \longrightarrow \left(\begin{matrix} 6 & 8 \\ 8 & 5 \end{matrix} \right).$$

This proves that the inverse of πb is $\pi^{-1}b$, and completes the proof of Theorem 7.3.

∎

COROLLARY 7.4 *The number of standard tableaux with* n *entries is equal to the number of involutions* π *of* [n].

Proof An *involution* of [n] is a permutation π of [n] such that $\pi = \pi^{-1}$. By Theorem 7.3, the Schensted correspondence is a bijection between all pairs of standard tableaux (P, P), and all permutations π of [n] such that $\pi = \pi^{-1}$.

The third application involves the matching f in the Boolean algebra of §2.2. We shall use the generalized Schensted correspondence to obtain the matching f.

Let $A \subset [n]$, $|A| = p$, and write A as an n-tuple of $n - p$ 0's and p 1's. Written this way, we may consider A as a multiset permutation of p 1's and $n - p$ 0's. Now apply Algorithm 19 to A to obtain (P, Q). Since P is column strict with $n - p$ 0's and p 1's, P must have either one or two rows. If there is no 1 below the last 0 in row 1, change that 0 to a 1 to obtain a new tableau P'. (For $p < \lfloor n/2 \rfloor$ this will always be the case.) Next apply Algorithm 20 to (P', Q) to obtain a string of $p + 1$ 1's and $n - p - 1$ 0's. This string corresponds to a $p + 1$ element subset B of [n]. Let $g(A) = B$.

For example, if $n = 8$ and $A = \{1, 4, 7, 8\}$, the binary string is 10010011. Algorithm 19 gives

$$P = \begin{array}{|c|c|c|c|c|} \hline 0 & 0 & 0 & 0 & 1 \\ \hline 1 & 1 & 1 \\ \cline{1-3} \end{array} \qquad Q = \begin{array}{|c|c|c|c|c|} \hline 1 & 2 & 3 & 5 & 6 \\ \hline 4 & 7 & 8 \\ \cline{1-3} \end{array}$$

Then switching the last 0 of P to a 1 gives

$$P' = \begin{array}{|c|c|c|c|c|} \hline 0 & 0 & 0 & 1 & 1 \\ \hline 1 & 1 & 1 \\ \hline \end{array}$$

Applying Algorithm 20 to P' and Q yields 11010011, which corresponds to $B = g(A) = \{1, 2, 4, 7, 8\}$. Note that $f(A) = g(A)$, where f is the matching from §2.2.

THEOREM 7.5 *If* g *is defined as above, then* $g = f$, *the matching in the Boolean algebra of* §2.2.

Proof We will show that the Schensted encoding algorithm, when applied to $f(A)$, yields (P', Q). By Theorem 6.2, this implies that $f(A) = g(A)$.

To do this, we need to find out how the peaks of the graph of A (as in §2.2) are related to the encoding algorithm. Let A_i be the initial segment of A of length i. Put $P_i(A) = Sch_P(A_i)$ and $Q_i(A) = Sch_Q(A_i)$. Let $h_i(A)$ be the height of the graph of A after the ith step. The following technical lemma is central to the proof.

LEMMA 7.6 *Suppose that* A *and* B *are subsets of* $[n]$ *and* $1 \le i < j \le n$, $j - i = 2k$. *Suppose also that* $h_i(A) = h_j(A)$ *and* $h_i(A) \ge h_m(A)$ *for* $i \le m \le j$. *Finally, suppose that* A *and* B *agree between* i *and* j, *and that the shapes of* $P_i(A)$ *and* $P_i(B)$ *are identical. Then*

(1) $P_j(A) = P_i(A)$ *with* k 0's *added to row* 1 *and* k 1's *added to row* 2,
(2) $P_j(B) = P_i(B)$ *with* k 0's *added to row* 1 *and* k 1's *added to row* 2, *and*
(3) *the part of* $Q_j(A)$ *which was added to* $Q_i(A)$ *is identical to the part of* $Q_j(B)$ *which was added to* $Q_i(B)$.

Note: the hypotheses on the heights just mean that there is a "valley" between the "peaks" i and j in the graphs of A or B. The conclusion of the lemma is that the P and Q tableaux of A and B are built identically (in a simple way) between these peaks.

Proof Let $P_i(A)$ be the following tableau.

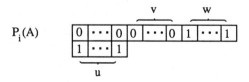

$P_i(A)$

Insertion of 0's increases the string (v) of 0's in row 1. Because $h_i(A) \geq h_m(A)$ for $i \leq m \leq j$, we have always added at least as many 0's as 1's. Thus, no 1 can bump a 1 in row 1, and all of the 1's are added to row 2. This proves (1). Because the shapes of $P_i(A)$ and $P_i(B)$ are identical, we also have shown (2).

By (1) and (2), inserting a 0 will create a new entry of Q in row 1, and inserting a 1 will create a new entry of Q in row 2. The same sequence of insertions is done for A and B because A and B agree between i and j. This proves (3).

∎

Lemma 7.6 means that if we are constructing P and Q for A and $B = f(A)$, we can ignore what happens between the peaks where A and B agree. For the graph of A, we replace all edges between such i and j by horizontal edges. For example, A = 110011101000100101011010010

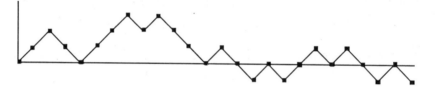

is replaced by:

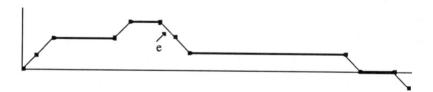

Generally, there will be a sequence of rising terraces followed by a sequence of falling terraces. The falling edge e of the highest terrace is the edge which is switched to construct f(A).

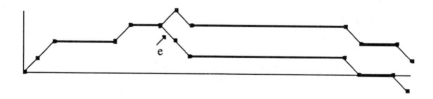

To prove Theorem 7.5, we need to show that $Sch_P(f(A)) = P'$ and $Sch_Q(f(A))$ $= Q$. Let the initial endpoint of the edge e have coordinates (i, j). Since A and $f(A)$ agree up to e, $P_i(A) = P_i(f(A))$ and $Q_i(A) = Q_i(f(A))$. Lemma 7.6 implies that $P_i(A)$ is the following tableau.

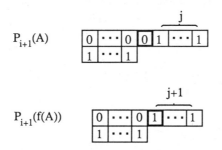

We insert a 0 into $P_i(A)$ to obtain $P_{i+1}(A)$, and 1 into $P_i(A)$ to obtain $P_{i+1}(f(A))$.

The cell in which $P_{i+1}(A)$ and $P_{i+1}(f(A))$ do not agree has been outlined. The new cell created in $P_{i+1}(A)$ and $P_{i+1}(f(A))$ is at the end of the first row, so $Q_{i+1}(A) = Q_{i+1}(f(A))$.

What happens as we continue to insert the remaining 0's and 1's of A? (Since $f(A)$ agrees with A past e, we are inserting the same 0's and 1's as in $f(A)$.) If we insert the 0's after e which are before the next terrace, they are placed in the first row of the P-tableau of A and $f(A)$. Again the only difference in these two tableaux will be the outlined entry. If we insert the entries of a terrace, Lemma 7.6 implies that the P and Q tableaux of A and $f(A)$ are built identically. Thus, $Sch_P(A)$ and $Sch_P(f(A))$ differ in only the outlined entry. This was our definition of P', so $Sch_P(f(A)) = P'$. We also conclude that $Sch_Q(f(A)) = Q$, which completes the proof of Theorem 7.5.

Notes

Two good references for bijections are Berge [Be] and Lothaire [Lot]. It is an active area of research (see, for example, the papers of Foata, Viennot or Zeilberger). The result for n-cycles and trees appears in Berge. Goulden and Jackson [G-J] also contains bijections with particular attention paid to generating functions. Knuth [Kn] is a rich source for constructions on permutations and the Schensted correspondence. The Schensted correspondence was equivalent to a correspondence of Robinson, who stated his correspondence with no proof. Knuth generalized it to two line arrays as in Exercise 41 below. It is sometimes referred to as the Robinson-Schensted-Knuth correspondence. The result of Exercise 30 is due to MacMahon in 1913. Foata [Fo] gave a bijective proof. Andrews [An] is the standard source for partitions. Many of the relationships between tableaux and permutations are given in James and Kerber [Ja-K].

Exercises

1.[2]   Prove by a bijection that

$$C_n = \sum_{k=0}^{n-1} C_k C_{n-k-1}.$$

2.[1]   Use a bijection to show that the number of terminal vertices in a full binary tree is one more than the number of internal vertices.

3.[1]   Given the sequence of well-formed parentheses $(\,(\,(\,)\,(\,)\,)\,(\,)\,)\,(\,)\,(\,(\,)\,(\,)\,)$, construct the corresponding

(a) binary tree,

(b) ordered tree,

(c) standard tableau.

4.[3]   A function $f : [n] \to [n]$ called *monotone* if $f(x) \le f(y)$ whenever $x \le y$. Give a bijection which shows that the number of monotone functions $f : [n] \to [n]$ which satisfy $f(i) \le i$, $i \in [n]$, is C_n. How many elements does Young's lattice $\mathcal{Y}_\lambda$ have, $\lambda = (n-1, n-2, \ldots, 1)$?

5.[2]   A rooted labeled binary tree T is called *increasing* if x > y whenever x is
below y in T. Thus 1 is the root of T. Give a bijection which proves that the
number of increasing binary trees with n vertices is n!.

6.[3C]   (Viennot and Zeilberger [Z3]) From §3.1 we know that there is a bijection
between ordered trees on n + 1 vertices and full binary trees on 2n + 1 vertices.
Write a program to investigate the following statistics on these trees.

Ordered Trees: A *filament* of a rooted tree T is a maximal path from a terminal
vertex, not including the root, all of whose vertices have degree ≤ 2. The filaments of
the tree below have been circled.

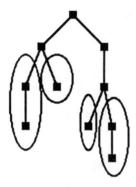

If the filaments of an ordered tree are deleted, another ordered tree results. The
filament number of an ordered tree T is the number of successive filament deletions
which reduce T to its root. The filament number of the above tree is 2. Here are the
5 ordered trees on 4 vertices and their filament numbers.

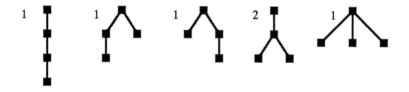

Full Binary Trees: The decomposition number of a full binary tree is defined
inductively. Label the terminal vertices of T with 0. Label any other vertex v of T
by the maximum of the labels of the two sons of v, if these labels are not the same,
and by the label +1 if these two labels are the same. The *decomposition number* of T
is the label of the root of T. Here are the 5 full binary trees on 7 vertices with their
appropriate labelings.

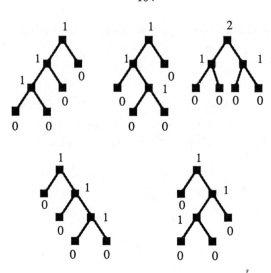

7.[3]   Using a bijection, prove

$$\sum_{k=0}^{n} \binom{n+k}{k} 2^{n-k} = 2^{2n}.$$

8.[1]   Find the Prüfer code of the following labeled tree.

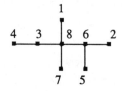

9.[2]   How many labeled trees on [n] have the degree sequence $d_1, d_2, \ldots, d_n$ where $d_i \geq 1$ and $d_1 + d_2 + \ldots + d_n = 2n - 2$?

10.[2]   How many labeled trees on n vertices are there such that vertex 1 has degree k?

11.[2]   Use Exercise 9 to find the generating function

$$F(x_1, \ldots, x_n) = \sum_{T} x_1^{d_1(T)} \cdots x_n^{d_n(T)},$$

where the sum is over all labeled trees T on [n], and $d_i(T)$ is the degree of vertex i in T?

12.[3]   Let $T_n = n^{n-2}$. By considering ordered pairs (T, e) of labeled trees T and edges e of T, give a bijective proof that

$$(n-1)\, T_n \; = \; \sum_{k=1}^{n-1} \binom{n-1}{k-1} T_k\, T_{n-k}\, k\, (n-k).$$

13.[1]   Let f be a function from [n] to [n], $f : [n] \to [n]$. The *functional digraph* of f is the directed graph G_f whose vertex set is [n] and with edge $i \to j$ if and only if $f(i) = j$. The graph G_π in the proof of Theorem 2.4 was a functional digraph. What does a typical functional digraph look like?

14.[2]   Find a bijection between all functional digraphs (see Exercise 13) on [n] with k loops (fixed points of f) and all function digraphs on [n] where 1 has in-degree k.

15.[2]   The following bijective proof of Cayley's theorem is due to G. Labelle [La]. A weighted version of it can be used to give a bijective proof of the *Lagrange inversion formula*.

Let S be the set of all (T, r, x), where T is a rooted labeled tree on n vertices, r is the root of T, and x is any vertex of T. Clearly, to prove Cayley's theorem it is sufficient to prove that $|S| = n^n$. So we need a bijection $\phi : S \to F$, where F is the set of all functions $f : [n] \to [n]$. We will give the functional digraph (see Exercise 13) of $f = \phi((T, r, x))$.

First, direct each edge of T toward the root r. Next, order the vertices of the unique path P in T from r to x. This gives a permutation of the vertices of P, whose cycles are the cycles of f. Attach the remaining vertices of T to these cycles in exactly the same way that they are attached in T, to complete the definition of f. Show that the map ϕ is a bijection.

Example: Let $n = 16$, $r = 4$, $x = 9$, and T be the tree (whose edges have already been directed)

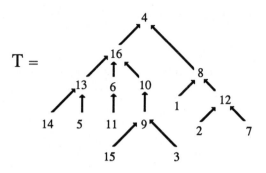

$$T =$$

Then the path P from r to x is 4 16 10 9, whose cycle decomposition is
(4) (9 16) (10). So we attach the rest of T to these cycles to define f.

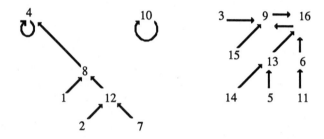

16.[3C]   Write a program to compute, for various values of m and n, the
generating function $G_{mn}(q)$ of §3.3. What do rank symmetry and unimodality of
Young's lattice y_λ for the shape $\lambda = n^m$ imply about $G_{mn}(q)$? Conjecture and
prove an explicit formula for $G_{mn}(q)$. (Hint: A useful polynomial is $(1-q)(1-q^2)$
$\cdots (1-q^n)$.) What justification can you give for $G_{mn}(q)$ being called the q-binomial
coefficient?

17.[2]   Prove that the number of partitions of n into k distinct parts is equal to the
number of partitions of $n-k(k+1)/2$ with at most k parts. What generating
function identity does this imply?

18.[2]   Use Theorem 4.1 to conclude that the inversion poset $\mathcal{J}_n$ is rank
symmetric and rank unimodal.

19.[1]   If $\pi = 372459168$, find the corresponding $\phi(\pi)$ of §3.4.

20.[2]   Prove by a bijection that $e(n, k) = e(n, n+1-k)$. What is the average

number of falls of a permutation π of n?

21.[3] Give a bijective proof of (4.4) for positive integral values of x. Then conclude that (4.4) holds as a polynomial identity in x.

22.[3] Prove by a bijection that

$$S(n, m) \; m! \; = \; \sum_{k=1}^{m} \; e(n, k) \; \binom{n-k}{m-k}.$$

23.[1] Prove the recurrence formulas (4.1) and (4.2) as outlined in §3.4.

24.[3] Can you find an orthogonality formula for the binomial coefficients?

25.[3] Can you find another orthogonality formula for the Stirling numbers?

26.[2] Prove (4.6) by using (4.5) and (4.3).

27.[3] Prove (4.4) from Exercise 22 above and Vandermonde's Theorem (Exercise 26 of Chapter 1).

28.[3C] Let $\tilde{e}(n, p, k)$ be the number of multiset permutations on $\mathfrak{M} = \{1^p, 2^p, \ldots, n^p\}$ that have k falls. For fixed values of n and p, write a program to find the values of $\tilde{e}(n, p, k)$. State and prove as many conjectures as you can. (This is called Simon Newcomb's problem.)

29.[4C] Let $\mathfrak{M}$ be the multiset of m 0's and n 1's. For $w \in \mathfrak{M}$, let $\phi^{-1}(w)$ be the corresponding partition of Theorem 3.2. Show that $\phi^{-1}(w)$ partitions the number inv(w). Write a program to find the generating function for the inversion number of a multiset of m 0's, n 1's, and k 2's. State and prove as many conjectures as you can.

✓ 30.[3C] (MacMahon, 1913; Foata [Fo]) The *index* of a permutation $\pi = \pi_1 \pi_2 \cdots \pi_n$ is the sum of all subscripts j such that $\pi_j > \pi_{j+1}$, $1 \le j \le n-1$. For example, the index of 47816325 is $3+5+6 = 14$. Write a program which finds, for a given n, the number of permutations a(n, k) which have index k. State and prove your conjectures. What is the generating function

$$\sum_{k=0}^{\infty} a(n, k)\, q^k \ ?$$

31.[1]   Give the Kostka table for $n = 4$.

32.[2]   Prove the Gale-Ryser Theorem (see, for example, [Ja-K]): $K_{\lambda \mu} > 0$ if and only if $\lambda \geq \mu$ in the dominance lattice.

33.[4]   Prove that $K_{\lambda \mu} \geq K_{\lambda \rho}$ if $\rho \geq \mu$ in the dominance lattice. (Hint: characterize $\rho \cdot > \mu$, and then show that $K_{\lambda \mu} \geq K_{\lambda \rho}$ if $\rho \cdot > \mu$.)

34.[4C]   Write a program which finds the number of column strict tableaux of shape λ whose entries are $\leq N$. State and prove your conjectures.

35.[1]   Suppose that $\pi = 443511242$. Find the P and Q tableaux of π in the generalized Schensted correspondence.

36.[1]   Suppose

$$P = \begin{array}{|c|c|c|c|}\hline 2 & 2 & 4 & 4 \\\hline 3 & 4 \\\cline{1-2} 4 & 5 \\\cline{1-2} 6 \\\cline{1-1}\end{array} \qquad , \qquad Q = \begin{array}{|c|c|c|c|}\hline 1 & 4 & 5 & 9 \\\hline 2 & 6 \\\cline{1-2} 3 & 7 \\\cline{1-2} 8 \\\cline{1-1}\end{array} \ .$$

Find the multiset permutation π which corresponds to (P, Q).

37.[2]   From the Schensted correspondence prove the Erdös-Szekeres Theorem: any sequence of $n^2 + 1$ distinct real numbers has either an increasing subsequence or decreasing subsequence of length $\geq n + 1$.

38.[3C]   Use Algorithm 19 to list all involutions of $[n]$ and the corresponding tableaux. Is there any relationship between the shape of the tableau and the cycle structure of the permutation?

39.[2]   A *lattice permutation* is a multiset permutation π of λ_1 1's, λ_2 2's, ... , λ_n n's, such that for any initial segment of π, the number of 1's $\geq$ the number of 2's $\geq \ldots \geq$ the number of n's. Thus, 1121233213 is a lattice permutation while

11212331321 is not. Find a bijection between all such lattice permutations and standard tableaux of shape λ.

40.[3C] (Schützenberger [Sc]) It is clear that a *row insertion algorithm* exists which is completely analogous to the column insertion algorithm. Write a program to investigate this problem: if π corresponds to (P, Q) under column insertion, and π' corresponds to (P, Q) under row insertion, how are π and π' related?

41.[3C] (Knuth [Kn1]) The Schensted correspondence may be generalized to give a bijection between all pairs (P, Q) of column strict tableaux of the same shape and contents (μ, ρ), and integral matrices A whose row sums are μ and column sums are ρ. We define a two line array π from A by using A_{ij} pairs (i, j) in π. We order these pairs in π to be increasing in i, and for a fixed i, decreasing in j. Then we apply the column insertion algorithm to π (using the top line of π for the entries of Q). For example, if

$$\mu = 554$$

$$A = \begin{bmatrix} 2 & 2 & 1 \\ 3 & 1 & 1 \\ 1 & 2 & 1 \end{bmatrix} \begin{matrix} 5 \\ 5 \\ 4 \end{matrix}$$

$$\rho = 653 \quad 6 \quad 5 \quad 3$$

then π has 2 pairs $(1, 1)$, 2 pairs $(1, 2)$, 1 pair $(1, 3)$, etc., so

$$\pi = \begin{pmatrix} 1\,1\,1\,1\,1\,2\,2\,2\,2\,2\,3\,3\,3\,3 \\ 3\,2\,2\,1\,1\,3\,2\,1\,1\,1\,3\,2\,2\,1 \end{pmatrix}$$

and

$$P = \begin{array}{|c|c|c|c|c|c|c|c|c|} \hline 1 & 1 & 1 & 1 & 1 & 1 & 2 & 2 & 3 \\ \hline 2 & 2 & 2 & 3 \\ \cline{1-4} 3 \\ \cline{1-1} \end{array} \quad , \quad Q = \begin{array}{|c|c|c|c|c|c|c|c|c|} \hline 1 & 1 & 1 & 1 & 1 & 2 & 2 & 2 & 3 \\ \hline 2 & 2 & 3 & 3 \\ \cline{1-4} 3 \\ \cline{1-1} \end{array} \; .$$

If A corresponds to (P, Q), what matrix corresponds to (Q, P)? Also, give an expression for the number of such matrices (with a general μ and ρ) which involves the Kostka numbers.

CHAPTER 4

Involutions

Many combinatorial formulas include positive and negative values. It might first seem that a bijection is not the proper tool for dealing with these formulas. This is not the case, however. Such formulas can sometimes be proved by using an involution on a *signed set*. In fact, involutions may be used to prove theorems seemingly unrelated to combinatorics. This will be done in Section 3 for the Cayley-Hamilton Theorem.

A *signed set* A is a set which has been partitioned into two subsets, A^+ and A^- with $A^+ \cup A^- = A$ and $A^+ \cap A^- = \emptyset$. The elements of A^+ and A^- are called *positive* and *negative*, respectively. We are interested in the value $\|A\| = |A^+| - |A^-|$.

If some of the elements of A^- can be paired with some of the elements of A^+, then the total size of the sets that we have to count to compute $\|A\|$ is reduced. In fact, if A^+ is bigger than A^- and if we pair up all of A^-, then $\|A\|$ is just the number of elements of A^+ which are unpaired. More formally, such a matching is an *involution* φ on A, that is, a permutation φ on A such that $\varphi^2 = \text{id}$. This involution has the property that whenever $\varphi(x) \neq x$, then $x \in A^+$ if and only if $\varphi(x) \in A^-$. Note that this means that if $x \in A^-$ then $\varphi(x) \in A^+$.

Notice that $\|A\|$ is precisely the number of fixed points of φ in A^+ minus the number of fixed points of φ in A^-. If we write $F(\varphi)$ for the fixed point set of φ, and $F(\varphi)^+ = F(\varphi) \cap A^+$ and $F(\varphi)^- = F(\varphi) \cap A^-$, then $F(\varphi)$ is a new signed set and $\|F(\varphi)\| = \|A\|$. Typically, one or both of $F(\varphi)^+$ or $F(\varphi)^-$ is empty. Such an involution φ is called *sign-reversing*, for if x is not fixed by φ, then $\varphi(x)$ has the sign opposite from x.

An important formula from elementary enumeration theory is the *principle of inclusion-exclusion*. This principle can easily be proved with a sign-reversing involution. Suppose X is some finite set of objects and each of these objects is endowed with certain *properties*. A property may be thought of as a subset of X. Suppose P denotes the collection of properties. So associated with $x \in X$ there is a subset $P_x \subset P$ of properties with which x is endowed. For any subset $T \subset P$ of properties let $N_=(T) = \{x \in X : P_x = T\}$ and $N_\supset(T) = \{x \in X : P_x \supset T\}$. Thus,

$N_=(T)$ is the set of objects in X whose list of properties is exactly T, while $N_{\supset}(T)$ is the set whose list contains T. The principle of inclusion-exclusion can now be stated:

$$(0.1) \qquad N_=(\varnothing) \;=\; \sum_{T \subset P} (-1)^{|T|} \, N_{\supset}(T).$$

Equation (0.1) is usually expressed in basic combinatorics texts as an alternating sum of various set unions and is generally proved by induction. In the form given here, it is quite easy to prove using an involution [Z2]. Let $A = \{(x, T) : x \in X$ and $T \subset P_x\}$. This set can be made into a signed set by defining $sgn(x, T) = (-1)^{|T|}$. Then $A^+ = \{(x, T) \in A : sgn(x, T) = +1\}$ and $A^- = \{(x, T) : sgn(x, T) = -1\}$. The right-hand side of (0.1) clearly computes $\|A\|$.

We now give a sign-reversing involution φ on A. Suppose the properties P are linearly ordered. For $(x, T) \in A$, let t be the largest property in P_x. If $t \in T$, then $\varphi(x, T) = (x, T - \{t\})$. If $t \in P_x - T$, then $\varphi(x, T) = (x, T \cup \{t\})$. Since $|T|$ changes by one, φ is sign-reversing. Two applications of φ will clearly restore (x, T). This construction cannot be accomplished if P_x is empty. In this case, T must be empty and $sgn(x, T) = +1$. These elements of A are the fixed points of φ and are all positive. They are clearly counted by the left-hand side of (0.1). This completes the proof.

§4.1 The Euler Pentagonal Number Theorem

In Exercise 15 of Chapter 1 you were asked to discover a relationship between partitions of n into an odd number of distinct parts and partitions of n into an even number of distinct parts. This relationship is called Euler's pentagonal number theorem. In this section we give a famous classical proof of it due to Franklin, which uses a sign-reversing involution.

THEOREM 1.1 *Let* PDE(n) (PDO(n)) *be the set of partitions of* n *into distinct parts with an even (odd) number of parts. Then*

$$PDE(n) - PDO(n) \;=\; \begin{cases} 0 & \text{if } n \neq (3k^2 \pm k)/2 \\ (-1)^k & \text{if } n = (3k^2 \pm k)/2. \end{cases}$$

Note: It is easy to see that Theorem 1.1 is equivalent to

$$(1.1) \qquad \prod_{i=1}^{\infty} (1-x^i) = 1 + \sum_{k=1}^{\infty} (-1)^k x^{(3k^2 \pm k)/2},$$

which is a special case of the Jacobi triple product identity [An]. It is called the pentagonal number theorem because the numbers $(3k^2 \pm k)/2$ arise when constructing larger regular pentagons from smaller ones.

Proof Let PD(n) be the set of all partitions of n into distinct parts. Let $PD(n)^+ = $ PDE(n) and $PD(n)^- = PDO(n)$. This makes PD(n) into a signed set with $sgn(\lambda) = (-1)^{\# \text{ of parts in } \lambda}$. The idea of this proof will be to construct a sign-reversing involution φ on PD(n) with no fixed points, unless n is of the form $n = (3k^2 \pm k)/2$, in which case φ will have exactly one fixed point. The sign of this fixed point will be $(-1)^k$. Clearly, if φ has these properties, Theorem 1.1 follows.

Suppose $\lambda \in PD(n)$. Recall that we write $\lambda = (\lambda_1, \lambda_2, \dots)$ with $\lambda_1 > \lambda_2 > \dots$. (The inequalities are strict here because the parts of λ are <u>distinct</u>.) Let $a(\lambda) = \max\{j : \lambda_j = \lambda_1 + 1 - j\}$ and $b(\lambda) = \min\{\lambda_j\}$. Thus $b(\lambda)$ is the smallest part of λ and $a(\lambda)$ is the length of the "staircase" on the border of the Ferrers diagram of λ. For $\lambda = (7, 6, 5, 3, 2)$, $a(\lambda) = 3$ and $b(\lambda) = 2$.

If $b(\lambda) \le a(\lambda)$, create a new partition $\varphi(\lambda)$ of n by moving the $b(\lambda)$ part adjacent to $a(\lambda)$. Thus $\varphi(7, 6, 5, 3, 2) = (8, 7, 5, 3)$:

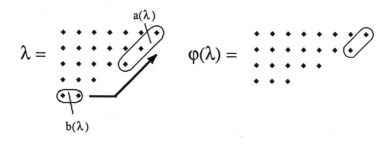

Note that since λ has distinct parts, $b(\varphi(\lambda)) > a(\varphi(\lambda)) = b(\lambda)$. The reader should carefully check this.

If $b(\lambda) > a(\lambda)$, then $\varphi(\lambda)$ is obtained by creating a "part" consisting of the $a(\lambda)$ cells at the end of the first $a(\lambda)$ rows. This part is placed under the $b(\lambda)$ part, i. e., this new part is now the smallest part of $\varphi(\lambda)$. For example, $\varphi(9, 8, 6, 3) = (8, 7, 6, 3, 2)$:

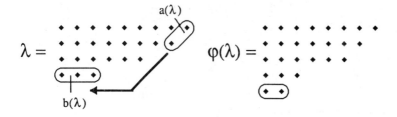

Since φ changes the number of parts by one, it is clear that it is sign-reversing. Furthermore, φ reverses the two cases above, i. e., if $b(\lambda) \leq a(\lambda)$ then $b(\varphi(\lambda)) > a(\varphi(\lambda))$ and conversely. But is φ defined on all of $PD(n)$? If it is not, we can extend φ by defining φ to <u>fix</u> these remaining partitions. Certainly φ <u>is</u> well-defined when the cells counted by $a(\lambda)$ and $b(\lambda)$ do not overlap. The reader should also check to see that it is well-defined if $a(\lambda) < b(\lambda) - 1$ or $a(\lambda) > b(\lambda)$, <u>whether or not these cells overlap</u>.

But if $b(\lambda) = a(\lambda)$ and these cells overlap, φ is not defined. Nor is it defined if $a(\lambda) = b(\lambda) - 1$ and these cells overlap. In the former case, if $b(\lambda) = k$, then $\lambda = (2k-1, 2k-2, \dots, k+1, k)$ so λ partitions $(3k^2 - k)/2$ and λ has k parts which means the sign of λ is $(-1)^k$. For example, λ might be $(7, 6, 5, 4)$ and $k = 4$:

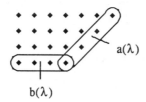

Moving the $b(\lambda)$ cells up will not give a proper partition.

In the latter case, if $b(\lambda) = k + 1$, then $\lambda = (2k, 2k-1, \dots, k+1)$, so λ is a partition of $(3k^2 + k)/2$, λ has k parts, and the sign of λ is $(-1)^k$. For example, λ might be $(6, 5, 4)$ and $k = 3$:

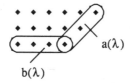

As an application of Theorem 1.1, recall that the generating function for all partitions is given by

$$(1.2) \qquad \prod_{i=1}^{\infty} (1-x^i)^{-1} \; = \; \sum_{n=0}^{\infty} p(n)\, x^n.$$

Clearing the fraction and substituting (1.1) gives

$$(1.3) \qquad \sum_{n=0}^{\infty} p(n)\, x^n \; \sum_{k=-\infty}^{\infty} (-1)^k x^{(3k^2+k)/2} \; = \; 1.$$

Equating coefficients in (1.3) gives, for $n > 0$,

$$(1.4) \qquad \sum_{k=-\infty}^{\infty} (-1)^k \, p(n - (3k^2 + k)/2) \; = \; 0.$$

For any given value of n, this sum is actually finite and gives a recurrence for $p(n)$.

§4.2 Vandermonde's Determinant

Sign-reversing involutions are a natural tool for handling identities involving determinants because the terms in the expansion automatically have signs attached. In this section we give a proof due to Gessel [Ge] which establishes Vandermonde's determinant using a sign-reversing involution.

Vandermonde's determinant is

$$(2.1) \qquad \det(\, x_j^{n-i}\,)_{1 \le i,j \le n} \; = \; \prod_{1 \le i < j \le n} (x_i - x_j).$$

It is clear that the product side of (2.1) has

$$2^{\binom{n}{2}}$$

terms, while the determinant has only $n!$ terms. We need a sign-reversing involution φ that cancels

$$2^{\binom{n}{2}} - n!$$

terms on the product side of (2.1). But first we must interpret the product as a generating function for an appropriate class of combinatorial objects.

A *tournament* T is a labeled directed graph on [n] such that any pair $\{i, j\}$, $i \neq j$, has exactly one directed edge, either $i \rightarrow j$ or $j \rightarrow i$. The word tournament is appropriate if we interpret $i \rightarrow j$ as "i beats j." Then each player i must play one game with each of the other $n - 1$ players. For each pair (i, j), $i < j$, there is a corresponding term, $x_i - x_j$, on the product side of (2.1). To each edge e of T assign a *weight* $w(e)$, with $w(e) = x_i$ if $e = i \rightarrow j$. That is, the weight of an edge is x subscripted by the winner. To each edge e of T we assign a *sign*, $sgn(e)$, with $sgn(e) = +1$ if $e = i \rightarrow j$ and $i < j$ and $sgn(e) = -1$ if $e = i \rightarrow j$ and $j < i$. The weight of a tournament T is defined by

$$(2.2) \qquad w(T) = \prod_{\text{edges } e} w(e).$$

It is clear that

$$(2.3) \qquad w(T) = x_1^{a_1} x_2^{a_2} \cdots x_n^{a_n},$$

where $a_i = $ the number of games player i wins. The sign of T, $sgn(T)$, is defined similarly. It is $(-1)^m$ where m is the number of ordered pairs (i, j) such that $i < j$ but j beats i. So we have shown that

$$(2.4) \qquad \prod_{1 \leq i < j \leq n} (x_i - x_j) = \sum_T w(T)\, sgn(T).$$

Of the

$$2^{\binom{n}{2}}$$

possible tournaments T, there are n! special tournaments: call T *transitive* if there is a *ranking* of players, $\pi_1 \pi_2 \cdots \pi_n$, such that π_i beats π_j if and only if $i < j$. Here is a transitive tournament T with ranking 35142:

$$T = \quad$$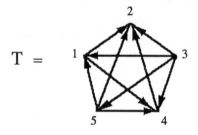

To prove (2.1), we need a sign-reversing involution φ on non-transitive tournaments which preserves the weight. If such a φ exists, (2.4) becomes

$$(2.5) \qquad \prod_{1 \le i < j \le n} (x_i - x_j) = \sum_{\substack{T \\ \text{transitive}}} w(T) \, \text{sgn}(T).$$

Any transitive T corresponds to a ranking permutation $\pi = \pi_1 \pi_2 \cdots \pi_n$, such that π_i wins $n - i$ games. For such T,

$$w(T) = x_{\pi_1}^{n-1} \cdots x_{\pi_n}^0.$$

Also, $\text{sgn}(T) = (-1)^m$, where m is the number of inversions of π. This is precisely the definition of the sign of the permutation π, so (2.5) becomes

$$(2.6) \qquad \prod_{1 \le i < j \le n} (x_i - x_j) = \sum_{\pi} \text{sgn}(\pi) \, x_{\pi_1}^{n-1} \cdots x_{\pi_n}^0.$$

But the right-hand side of (2.6) is the definition of the determinant in (2.1).

The proof then hinges on the involution φ. We need first a characterization of non-transitive tournaments that you are asked to show in Exercise 17.

PROPOSITION 2.1 T *is a non-transitive tournament if and only if* T *has two vertices with equal out-degree.*

Proof Exercise.

The sequence of out-degrees, or *wins*, $(a_1, a_2, \ldots, a_n)$ is called the *score vector* $\bar{a}$ of T. Choose the lexicographically first pair (i, j), $i < j$, such that $a_i = a_j$. For example, the tournament below has score vector $(2, 3, 0, 3, 2, 6, 5)$; choose $i = 1$ and $j = 5$.

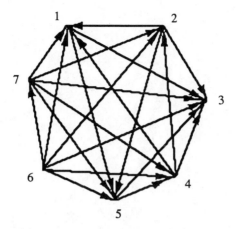

Assume, without loss of generality, that the directed edge between i and j is i → j. Consider all the other vertices k. The vertices i, j and k will form one of four kinds of triangles:

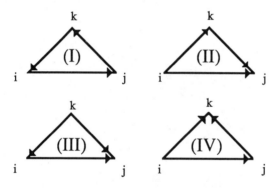

The tournament φ(T) is obtained from T by reversing all of the directed edges on triangles of types (I) and (II). At least one such triangle must exist because i and j have the same out-degrees, but the edge i → j contributes one to the out-degree of i and zero to the out-degree of j. In the example above, i = 1 and j = 5; the triangles 1-5-7, 1-5-6, and 1-5-2 are type (III); the triangle 1-5-3 is type (IV); and the triangle 1-5-4 is type (I). There are no type (II) triangles. These triangles are indicated in the drawing below; all other triangles have been omitted.

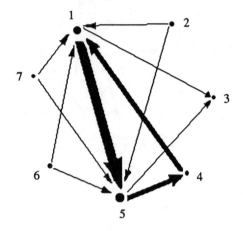

The involution φ reverses the edges of the 1-5-4 triangle:

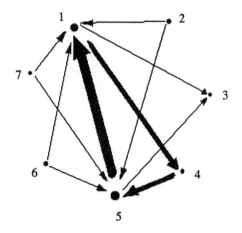

Thus φ produces this tournament:

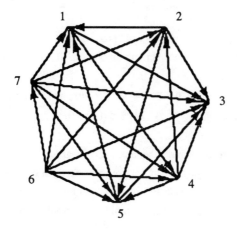

Certainly the out-degrees of all vertices $k \neq i$ or j are unchanged by φ. But what about a_i and a_j? Let $\hat{a}_i$ and $\hat{a}_j$ denote the out-degrees of i and j in $\varphi(T)$. Notice that the only contribution to a_i comes from triangles of type (II) and type (IV), plus the $i \rightarrow j$ edge, while the only contribution to a_j comes from triangles of type (I) and type (IV). But the only contribution to $\hat{a}_i$ is from triangles of type (I) and type (IV), while the only contribution to $\hat{a}_j$ is from triangles of type (II) and type (IV), plus the $i \rightarrow j$ edge. This means that $\hat{a}_i = a_j$ and $\hat{a}_j = a_i$. Since $a_i = a_j$, they are all equal.

This shows that

$$(2.7) \qquad w(T) = w(\varphi(T)).$$

Reversing an edge changes the sign of T. Since φ reverses two edges per type (I) or type (II) triangle and also reverses $i \rightarrow j$, it reverses an odd number of edges and so

$$(2.8) \qquad sgn(T) = - sgn(\varphi(T)).$$

It remains to show that φ is in fact an involution on non-transitive tournaments. Since φ fixes the score vector of T, the first pair $i < j$ in $\varphi(T)$ such that $a_i = a_j$ is the same as that pair for T. Finally, reversing the edges in types (I) and (II) preserves these types. So if φ is applied to $\varphi(T)$, the same edges are again reversed; thus $\varphi(\varphi(T)) = T$.

§4.3 The Cayley-Hamilton Theorem

In this section we give a combinatorial proof of the Cayley-Hamilton Theorem. This proof once again uses a sign-reversing involution. It is due to Straubing [Str].

THEOREM 3.1 *Let* A *be any* $n \times n$ *matrix over any field. Let* $p_A(\lambda) = \det(\lambda I - A)$ *be the characteristic polynomial of* A. *Then* $p_A(A) = 0$.

It might seem surprising that a theorem from linear algebra has a combinatorial proof. However, sign-reversing involutions are a perfectly suitable tool for handling determinants, because determinants are signed sums of products of the entries of a matrix. In fact, it can be argued that the combinatorial proof we give here is the most "natural" proof because it does not depend upon the field of scalars. Proofs of this theorem from algebra usually first prove a weak version for diagonal or triangular matrices and then "extend" to all matrices. However, this extension requires that the scalars be the complex numbers, and some major theorem, such as the Fundamental Theorem of Algebra or Taylor's Theorem, must be used to eliminate the dependence on the complex numbers.

Proof We begin by writing the characteristic polynomial as a signed sum of products

$$(3.1) \qquad p_A(\lambda) = \sum_{\pi \in S_n} \text{sgn}(\pi) \prod_{i=1}^{n} (\lambda I - A)_{i \, \pi(i)}.$$

Each fixed-point i of π (i. e., $\pi(i) = i$) will contribute either λ or $-a_{ii}$ to the product, while each non-fixed-point i will contribute $-a_{i\pi(i)}$. Equation (3.1) can now be written

$$(3.2) \qquad p_A(\lambda) = \sum_{\pi \in S_n} \text{sgn}(\pi) \sum_{S \subset [n]} (-1)^{|S|} \lambda^{n-|S|} \prod_{i \in S} a_{i \, \pi(i)},$$

where the subset $S \subset [n]$ is any subset which satisfies

$$(3.3) \qquad [n] - S \subset \text{fixed points of } \pi = F(\pi).$$

Note that since π fixes everything in [n] outside of S, we may regard π as a permutation of the elements of S. Let P(S) denote the set of permutations of S. The sign of π will be the same when it is regarded as a permutation in P(S). So we may reorder the sum in (3.2) and organize by $|S|$:

$$(3.4) \qquad p_A(\lambda) = \sum_{k=0}^{n} \lambda^{n-k} \sum_{|S|=k} \sum_{\pi \in P(S)} \mathrm{sgn}(\pi)\,(-1)^k \prod_{i \in S} a_{i\,\pi(i)}.$$

It is well-known that the sign of a permutation is the product of the signs of its cycles, and that the sign of a cycle of length r is $(-1)^{r-1}$. So each cycle in π contributes $(-1)^r$ to $(-1)^k$ and $(-1)^{r-1}$ to $\mathrm{sgn}(\pi)$. Thus the total contribution of a given cycle to $(-1)^k\,\mathrm{sgn}(\pi)$ is -1 and $(-1)^k\,\mathrm{sgn}(\pi)$ can be replaced with $(-1)^{d(\pi)}$, where $d(\pi)$ is the number of cycles in π

$$(3.5) \qquad p_A(\lambda) = \sum_{k=0}^{n} \lambda^{n-k} \sum_{|S|=k} \sum_{\pi \in P(S)} (-1)^{d(\pi)} \prod_{i \in S} a_{i\,\pi(i)}.$$

We can visualize a typical permutation $\pi \in P(S)$ as a directed graph on the vertices $[n]$, with edges $i \to \pi(i)$. This is the graph G_π we encountered in §3.2. In this graph, each vertex in S will have in-degree and out-degree equal to one. Also, each edge $i \to \pi(i)$ will be given a weight $a_{i\pi(i)}$; and each cycle π will correspond to a cycle in the graph and will be given the sign -1. Then the weight of π will be

$$w(\pi) = \prod_{i \in S} a_{i\,\pi(i)}$$

and the sign of π will no longer be the ordinary sign of a permutation, but will be $\mathrm{sgn}^*(\pi) = (-1)^{d(\pi)}$. In the example below, $n = 9$, $S = \{1, 4, 6, 9\}$ and $\pi = (146)\,(9)$. Then $w(\pi) = a_{14}a_{46}a_{61}a_{99}$ and $\mathrm{sgn}^*(\pi) = +1$.

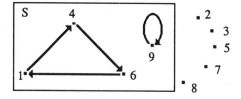

Next, replace λ in (3.5) with A. We wish to describe the ij-th entry of the resulting matrix.

$$(3.6) \qquad (p_A(A))_{ij} = \sum_{k=0}^{n} (A^{n-k})_{ij} \sum_{|S|=k} \sum_{\pi \in P(S)} (-1)^{d(\pi)} \prod_{t \in S} a_{t\pi(t)}.$$

This involves describing $(A^{n-k})_{ij}$. This is clearly

$$(3.7) \qquad (A^{n-k})_{ij} = \sum_{i_1, i_2, \ldots, i_{n-k-1}} a_{i\,i_1} a_{i_1 i_2} \cdots a_{i_{n-k-1}\, j}.$$

We visualize a term in this sum combinatorially as a directed path P of length $n - k$ from i to j on the vertices $[n]$. Moreover, let the weight of P, $w(P)$, be the product of the weights of the edges of P where the weight of an edge is as above: $w(e) = a_{ij}$ if $e = i \rightarrow j$. Then $w(P)$ is exactly a term in this sum.

We may now give a complete combinatorial description of the right hand side of (3.6). Let (S, π, P) be a triple such that

(a) S is a subset of $[n]$;

(b) π is a permutation on S; and

(c) P is a directed path from i to j of length $n - |S|$ with vertices in $[n]$.

The weight of the triple, $w(S, \pi, P)$, is $w(P)\, w(\pi)$ and the sign of the triple, $\text{sgn}(S, \pi, P)$, is $\text{sgn}^*(\pi)$. Then the ij-th entry of $p_A(A)$ is the generating function

$$(3.8) \qquad (p_A(A))_{ij} = \sum_{(S, \pi, P)} w(S, \pi, P)\, \text{sgn}(S, \pi, P).$$

Clearly, the set of such triples, S_{ij}, is a signed set. To prove the Cayley-Hamilton Theorem, we need to show that (3.8) is zero. Thus, we require a weight-preserving, sign-reversing involution φ on S_{ij}.

We can visualize a triple $(S, \pi, P) \in S_{ij}$ as a directed multi-graph on the vertices $[n]$ with two kinds of edges: the edges from π and the edges used in P. Note that the path P may use an edge more than once and may also use edges in π; hence the graph is a multi-graph. As an example, let $n = 9$, $i = 2$, $k = 4$, and $j = 5$. Choose $S = \{1, 4, 6, 9\}$, $\pi = (146)\,(9)$ and $P = 211515$.

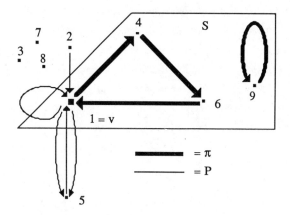

As another example with the same n, i, j, S and π, let P = 253215.

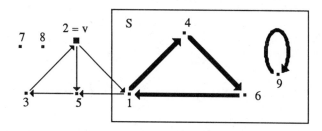

If φ is weight-preserving, it should preserve the edges of this multi-graph. If φ is sign-reversing, it should change the number of cycles in π by one. We look for such an involution.

Note that either the path P intersects π or P contains a cycle. For if P did not intersect π, it would be a graph on n − |S| vertices with n − |S| edges, and would therefore have a cycle.

Let v be the first vertex in P such that either

(1)   v is in π, or

(2)   v completes a cycle in P.

By the preceding discussion, at least one vertex satisfying (1) or (2) must exist. So we can choose the first such vertex. Furthermore, this vertex cannot satisfy <u>both</u> (1) and (2); for if v is in π and completes a cycle in P, it was encountered in P before and would have satisfied (1) at that point. In the first example above, v = 1 and is chosen by (1); in the second example, v = 2 and is chosen by (2).

We can now define φ(S, π, P) = (S̃, π̃, P̃). Suppose v satisfies (1); let C be the cycle of π containing v. Let $\tilde{P}$ be P with C inserted at the position of v. Let $\tilde{S}$ be S with the vertices in C removed. Let $\tilde{\pi}$ be π with C removed. Note that the second occurrence of v in $\tilde{P}$ now completes a cycle in $\tilde{P}$. No earlier vertex in $\tilde{P}$ satisfies (2): those up to and including the first occurrence of v are the same as in P and they did not satisfy (2) in P; those between the two occurrences of v were on C and therefore could not have been encountered before v in P. Therefore $(\tilde{S}, \tilde{\pi}, \tilde{P})$ satisfies (2). In our first example, $\tilde{S} = \{9\}$ and $\tilde{P} = 214611515$:

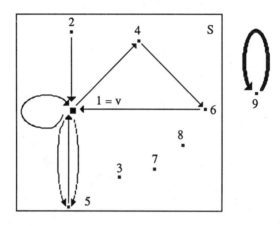

Now suppose v satisfies (2); let C be the cycle in P just completed. Then C will be a cycle from v to v in P. Let $\tilde{P}$ be P with C removed (including one occurrence of v). Let $\tilde{S}$ be S with the vertices in C added. Let $\tilde{\pi}$ be π with the cycle C added. This construction is legal because no vertex before the second occurrence of v could be in π or could be a repetition in P. Note that the first occurrence of v in $\tilde{P}$ now satisfies (1) because v is now in π. In our second example, $\tilde{S} = \{1, 2, 3, 4, 5, 6, 9\}$ and $\tilde{P} = 215$:

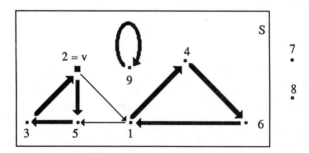

Since φ sends triples satisfying (1) to triples satisfying (2) and vice versa, φ is an involution. Furthermore, φ either removes or adds one cycle to π, so it is sign-reversing. Finally, all the edges of the graph of the triple remain, so it is weight-preserving.

§4.4. The Matrix-Tree Theorem

Suppose G is a labeled graph. A *spanning tree* of G is a tree with the same vertex set as G and with an edge set that is a subset of the edges set of G. We may then ask how many spanning trees does G have? The answer is given by the matrix-tree theorem.

THEOREM 4.1 *Let* G *be a labeled graph on* [n]. *Then the number of spanning trees of* G *is any cofactor of the* $n \times n$ *matrix* $D - A$, *where* D *is diagonal,* $(D)_{ii} = d_i$, *the degree of vertex* i, *and* $(A)_{ij} = 1$ *if* $i \text{---} j$, $i \neq j$, *is an edge of* G *and* $(A)_{ij} = 0$ *otherwise.*

For example, if $G = K_3$, the complete graph on 3 vertices, then

$$(4.1) \qquad D - A = \begin{bmatrix} 2 & -1 & -1 \\ -1 & 2 & -1 \\ -1 & -1 & 2 \end{bmatrix}$$

and any cofactor of $D - A$ has the value 3. So K_3 has 3 spanning trees.

If $G = K_n$, Theorem 4.1 gives n^{n-2} as the number of spanning trees (see Exercise 22). So Theorem 4.1 gives another proof of Cayley's Theorem (Theorem 2.1 in Chapter 3).

In this section, we prove Theorem 4.1 using a sign-reversing involution φ. This proof is due to Chaiken [Ch] and is very similar to the proof of the previous section. However, this time φ will have a fixed point set.

Proof It is more convenient to work with rooted labeled trees. Let $\mathfrak{I}_n$ be the rooted trees on [n] with root n. Such trees will be considered directed, with each edge directed toward the root. Suppose $T \in \mathfrak{I}_n$. For each directed edge $e = i \rightarrow j$ of T, assign the weight $w(e) = a_{ij}$. As in the previous section, $w(T)$ is the product of the weights of edges in T. For example, if $n = 7$ and

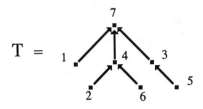

$$T = $$

then $w(T) = a_{17}a_{47}a_{37}a_{24}a_{64}a_{53}$. Thus, $w(T)$ is a monomial in the variables a_{ij}, $1 \le i \ne j \le n$.

Let $\overline{a}$ denote the variables a_{ij}, $1 \le i, j \le n$. The generating function for the trees in $\mathfrak{I}_n$ is given by

(4.2) $f_n(\overline{a}) = \sum_{T \in \mathfrak{I}_n} w(T)$.

To count the number of spanning trees of G, merely put $a_{ij} = a_{ji} = 1$ for each pair (i, j) that is an edge in G and $a_{ij} = a_{ji} = 0$ for each pair (i, j) that is not an edge in G.

Let us now identify $f_{n+1}(\overline{a})$ as an $n \times n$ determinant.

PROPOSITION 4.2 Let $R_i = a_{i1} + a_{i2} + \ldots + a_{i\,n+1}$, $1 \le i \le n$. Then $f_{n+1}(\overline{a}) = \det(R_i \delta_{ij} - a_{ij})$, $1 \le i, j \le n$.

Note: For $n = 2$, Proposition 4.2 is

(4.3) $f_3(\overline{a}) = \det \begin{bmatrix} a_{12} + a_{13} & -a_{12} \\ -a_{21} & a_{21} + a_{23} \end{bmatrix} = a_{12}a_{23} + a_{13}a_{21} + a_{13}a_{23}$,

which is the generating function for the three rooted labeled trees on [3].

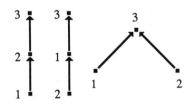

Clearly, Proposition 4.2 implies the matrix-tree theorem for the $(n+1, n+1)$ cofactor and thus for the (i, i) cofactor. It can be adapted for the (i, j) cofactor, but we do not do so here.

Proof of Proposition 4.2 Proceeding as in the derivation of (3.5) in §4.3, we get

$$(4.4) \qquad \det (R_i \delta_{ij} - a_{ij}) = \sum_{S \subset [n]} \sum_{\pi \in P(S)} (-1)^{d(\pi)} w(\pi) \prod_{i \in [n]-S} R_i \, ,$$

where S is a subset of $[n]$, π is a permutation of S with $d(\pi)$ cycles, and the weight of π, $w(\pi)$, is

$$(4.5) \qquad w(\pi) = \prod_{i \in S} a_{i \, \pi(i)} \, .$$

The only difference between (4.4) and (3.5) is that

$$\prod_{i \in [n]-S} R_i$$

replaces $\lambda^{n-|S|}$. Expanding this product gives

$$(4.6) \qquad \prod_{i \in [n]-S} R_i = \sum_{f:[n]-S \to [n+1]} w(f),$$

where

$$(4.7) \qquad w(f) = \prod_{i \in [n]-S} a_{i \, f(i)} \, .$$

So

$$(4.8) \qquad \det (R_i \delta_{ij} - a_{ij}) = \sum_{S \subset [n]} \sum_{\pi \in P(S)} \sum_{f:[n]-S \to [n+1]} (-1)^{d(\pi)} w(\pi) w(f).$$

The combinatorial description of the right-hand side of (4.8) is the set of triples (S, π, f) such that

(a) S is a subset of $[n]$;

(b) π is a permutation on S; and

(c) f is a function from $[n] - S$ to $[n+1]$.

The weight of a triple, $w(S, \pi, f)$, is $w(f)\,w(\pi)$; the sign of a triple, $\mathrm{sgn}(S, \pi, f)$, is $\mathrm{sgn}^*(\pi) = (-1)^{d(\pi)}$, as in §4.3. To prove Proposition 4.2, we must find a weight-preserving, sign-reversing involution φ on these triples.

Such a triple can be represented as a directed graph on $[n+1]$ with two kinds of edges: those which represent π, as in §4.3; and those which represent f, that is, an edge $i \to j$ if $f(i) = j$, or the functional digraph of f (see Exercise 13 of Chapter 3). For example, if $n = 10$, $S = \{3, 5, 7, 8, 9\}$, $\pi = (398)\,(5)\,(7)$ and

$$f = \begin{pmatrix} 1 & 2 & 4 & 6 & 10 \\ 11 & 4 & 2 & 3 & 7 \end{pmatrix},$$

then this triple can be represented:

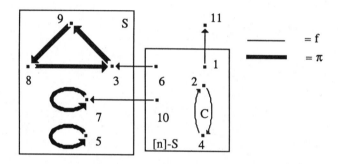

As another example, let $n = 10$, $S = \{3, 5, 7, 8, 9\}$, $\pi = (398)\,(5)\,(7)$ and

$$f = \begin{pmatrix} 1 & 2 & 4 & 6 & 10 \\ 4 & 1 & 11 & 8 & 10 \end{pmatrix}.$$

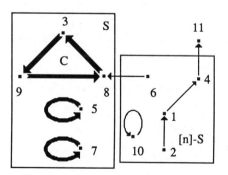

Once again, φ will change the number of cycles of π by exactly one and will preserve all the edges in the graph. Let C be the cycle in this graph with the smallest vertex. Define φ(S, π, f) = (S̃, π̃, f̃) as follows. If C is in π, let S̃ be S with C removed, f̃ be f with C added, and π̃ be π with C removed. If C is in f, let S̃ be S with C added, f̃ be f with C removed, and π̃ be π with C added. In the first example, S̃ = {2, 3, 4, 5, 7, 8, 9}, π̃ = (24) (398) (5) (7) and

$$\tilde{f} = \begin{pmatrix} 1 & 6 & 10 \\ 11 & 3 & 7 \end{pmatrix}$$

to give the graph:

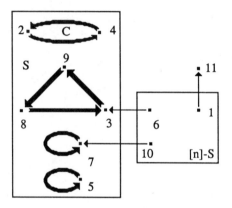

In the second example, S̃ = {5, 7}, π̃ = (5) (7) and

$$\tilde{f} = \begin{pmatrix} 1 & 2 & 3 & 4 & 6 & 8 & 9 & 10 \\ 4 & 1 & 9 & 11 & 8 & 3 & 8 & 10 \end{pmatrix}$$

to give the graph:

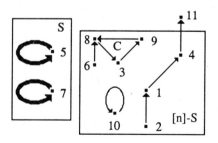

Clearly, φ is sign-reversing and weight-preserving; also $\varphi^2 = \text{id}$ because C will move back to its original position. What are the fixed-points of φ? They will be those triples (S, π, f) with no cycles. Thus, π is empty, $S = \varnothing$, and the graph of $f : [n] \to [n+1]$ has no cycles. This graph must then be a tree, with every edge directed along the path toward $n + 1$, that is, a tree in $\mathfrak{J}_{n+1}$. Conversely, any tree in $\mathfrak{J}_{n+1}$ naturally defines such a function f. The example at the beginning of this section corresponds to the function

$$f = \begin{pmatrix} 1 & 2 & 3 & 4 & 5 & 6 \\ 7 & 4 & 7 & 7 & 3 & 4 \end{pmatrix}.$$

So the right-hand side of (4.8) becomes

$$(4.9) \qquad \sum_{T \in \mathfrak{J}_{n+1}} w(T) = f_{n+1}(\overline{a}),$$

which is exactly what Proposition 4.2 claimed.

§4.5 Lattice Paths

Franklin's proof of the pentagonal number theorem appeared in 1881. Another early sign-reversing involution, called the reflection principle, was given by Andre in 1887. This time the signed set $S = S^+ \cup S^-$ consists of certain lattice paths in the plane. The involution reflects a lattice path through a line to obtain a new lattice path. In this section we use the reflection principle for the generalized ballot problem. We also use a related idea (due to Gessel and Viennot) to relate determinants of binomial coefficients to non-intersecting lattice paths and column strict tableaux. We prove a formula of Frobenius, which is a precursor of the hook formula for standard tableaux (Theorem 5.4 of Chapter 3).

In Chapter 3, we saw that the Catalan number C_n was the solution to the ballot problem: if candidates A and B both receive n votes, how many ways are there to count the votes so that A is never behind B? We give an alternative proof here, which uses the *reflection principle*. Represent any sequence of 2n votes as a lattice path (up for A, down for B) with unit steps, beginning at the origin. For example, ABAABABBAAABBABB is represented by

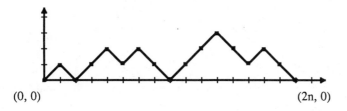

(0, 0)   (2n, 0)

Clearly there is a bijection between all sequences of $2n$ votes and such lattice paths. Moreover, if A and B have the same number of votes, n, the lattice path ends at $(2n, 0)$. The condition that A is never behind B is equivalent to the lattice path remaining at or above the x-axis.

Before defining the signed set $S = S^+ \cup S^-$, we make a minor change in the lattice paths we want to count. Any lattice path on or above the x-axis can be displaced vertically by one unit to a lattice path which lies <u>strictly</u> above the x-axis. That means there is a bijection between solutions to the ballot problem with $2n$ votes and lattice paths from $(0, 1)$ to $(2n, 1)$ which lie strictly above the x-axis.

Now let S^+ be the set of all lattice paths from $(0, 1)$ to $(2n, 1)$. Let S^- be the set of all lattice paths from $(0, -1)$ to $(2n, 1)$. Clearly, $|S^+|$ is

$$\binom{2n}{n}$$

while $|S^-|$ is

$$\binom{2n}{n+1}.$$

We will define a sign-reversing involution φ on $S^+ \cup S^-$ whose fixed point set $F(\varphi)$ consists exactly of the solutions to the ballot problem, all of which are in S^+. Thus,

(5.1) $|F(\varphi)| = \|S\| = |S^+| - |S^-| = C_n$.

To define φ, note that any path P in S^- <u>must</u> cross the x-axis. Let m be the smallest x-coordinate such that P crosses the x-axis at m. Reflect the initial segment of P from $x = 0$ to $x = m$ across the x-axis to obtain the new path $\varphi(P)$. Note that $\varphi(P) \in S^+$. Clearly φ is similarly defined for all $P \in S^+$ which touch or cross the x-axis and $\varphi^2 = \text{id}$. The fixed points of φ are those paths P in S^+ which do not touch or cross the x-axis.

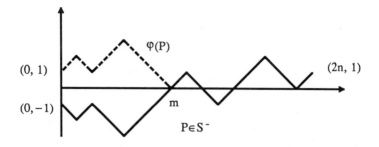

It is clear that this principle works for a more general ballot problem. Suppose A and B receive n and m votes respectively, $n \geq m$. How many ways are there to count the votes so that A is never behind B? This time we consider lattice paths from $(0, 1)$ to $(n+m, n-m+1)$. The reflection principle gives us the answer immediately:

$$\binom{n+m}{n} - \binom{n+m}{n+1} = \frac{n+1-m}{n+1}\binom{n+m}{n}.$$

The reflection principle was generalized by Gessel and Viennot [Ge-V] to allow k-tuples of lattice paths. They showed that there are many relationships between lattice paths, determinants and tableaux. We will present a few of these.

It is more convenient if we "tilt" our pictures 45°. A lattice path P will no longer consist of steps $(1, 1)$ (up) and $(1, -1)$ (down), but horizontal $(1, 0)$ and vertical $(0, 1)$ steps. For example, a lattice path P from $(1, 1)$ to $(4, 3)$ could be:

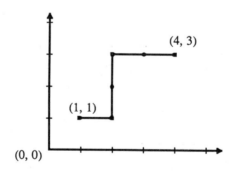

It is easy to see that such lattice paths are equivalent to the up-down lattice paths of the ballot problem. In fact, a solution to the ballot problem corresponds to a lattice path P from $(0, 0)$ to (n, n) which always lies at or above the line $y = x$.

The lattice paths we will consider always begin on the line $y = 1$. Let us write $P : (a, 1) \rightarrow (b, N)$ to mean a lattice path from $(a, 1)$ to (b, N). How many P are there such that $P : (a, 1) \rightarrow (b, N)$? Clearly, this number is

$$\binom{N+b-a-1}{b-a}.$$

Now fix integers $0 < a_1 < a_2 < \ldots < a_k$ and $0 < b_1 < b_2 < \ldots < b_k$ and N, and let S be the set of pairs $(\sigma, (P_1, P_2, \ldots, P_k))$ where σ is a permutation of $[k]$ and $(P_1, P_2, \ldots, P_k)$ is a k-tuple of lattice paths such that $P_i : (a_i, 1) \to (b_{\sigma(i)}, N)$. For example, let $k = 4$, $(a_1, a_2, a_3, a_4) = (1, 3, 4, 5)$, $(b_1, b_2, b_3, b_4) = (3, 5, 6, 7)$, $N = 5$ and $\sigma = (1)\,(2)\,(34)$.

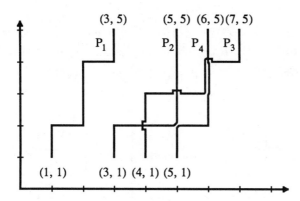

How many such k-tuples are there such that <u>none</u> of the lattice paths intersect?

THEOREM 5.1 *The number of k-tuples of lattice paths* $(P_1, P_2, \ldots, P_k)$ *such that*

(i) $P_i : (a_i, 1) \to (b_i, N)$, $1 \le i \le k$, *and*

(ii)   *any two paths* P_i *and* P_j *do not intersect*

is det M, *where* $M = (M_{ij})$ *is a* $k \times k$ *matrix with*

$$M_{ij} = \binom{N + b_j - a_i - 1}{b_j - a_i}.$$

Proof Let's begin with a simple example: $a_1 = 3$, $a_2 = 4$, $b_1 = 4$, $b_2 = 6$ and $N = 4$. Then Theorem 5.1 says there are

$$\binom{4}{1}\binom{5}{2} - \binom{3}{0}\binom{6}{3} = 20$$

such pairs (P_1, P_2). This special case is easy to prove and tells us how to proceed in general. There are certainly

$$\binom{4}{1}\binom{5}{2}$$

pairs of paths (P_1, P_2) such that $P_1 : (3, 1) \to (4, 4)$ and $P_2 : (4, 1) \to (6, 4)$. From these pairs we must exclude all pairs (there will be

$$\binom{3}{0}\binom{6}{3}$$

of them) (P_1, P_2) which intersect. If P_1 and P_2 intersect, let m be the last point of intersection. Construct $(\tilde{P}_1, \tilde{P}_2) = \varphi(P_1, P_2)$ by interchanging the paths from m to the endpoints. Then $\tilde{P}_1 : (3, 1) \to (6, 4)$ and $\tilde{P}_2 : (4, 1) \to (4, 4)$.

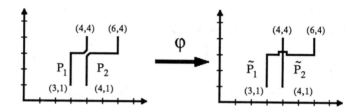

Any such pair $(\tilde{P}_1, \tilde{P}_2)$ must intersect so φ is a bijection to all such pairs. There are clearly

$$\binom{3}{0}\binom{6}{3}$$

pairs $(\tilde{P}_1, \tilde{P}_2)$.

Recall that S is the set of k-tuples of paths $(P_1, P_2, \dots, P_k)$, together with the permutation σ. The sign of σ makes S into a signed set. Clearly,

$$(5.2) \qquad \|S\| = |S^+| - |S^-| = \sum_{\sigma \in S_k} \text{sgn}(\sigma) \prod_{i=1}^{k} \binom{N + b_{\sigma(i)} - a_i - 1}{b_{\sigma(i)} - a_i},$$

which is det M in Theorem 5.1. All that is necessary is a sign-reversing involution φ on S whose fixed point set $F(\varphi)$ is given by (i) and (ii) of Theorem 5.1.

The "bad" elements of S are those with paths which intersect. These are the ones on which we must define φ. Let $(\sigma, (P_1, \dots, P_k)) \in S$ be such that at least two paths of $(P_1, \dots, P_k)$ intersect. Choose the first pair $i < j$ in lex order such that

P_i intersects P_j. Construct new paths $\tilde{P}_i$ and $\tilde{P}_j$ as before by switching the tails after the last point of intersection of P_i and P_j. Now the path $\tilde{P}_i$ will end at $(b_{\sigma(j)}, N)$ and $\tilde{P}_j$ will end at $(b_{\sigma(i)}, N)$. Since $\sigma_0(i\,j)$ sends i to $\sigma(j)$, j to $\sigma(i)$, and t to $\sigma(t)$ for $t \neq i, j$, we let

$$\varphi(\sigma, (P_1, \dots, P_k)) = (\sigma_0(i\,j), (P_1, \dots, \tilde{P}_i, \dots, \tilde{P}_j, \dots, P_k)).$$

Clearly φ is now sign-reversing. Since the first intersecting pair $i < j$ is not affected by φ, φ is an involution. The only paths $(P_1, \dots, P_k)$ which do not intersect have $\sigma =$ the identity.

Consider the example given earlier where $k = 4$, $(a_1, a_2, a_3, a_4) = (1, 3, 4, 5)$, $(b_1, b_2, b_3, b_4) = (3, 5, 6, 7)$, $N = 5$ and $\sigma = (1)\,(2)\,(34)$. The first pair $i < j$ of intersecting paths occurs when $i = 2$ and $j = 3$. The last point of intersection of P_i and P_j is at $(5, 3)$. The new 4-tuple of paths is given below. The new permutation will be $(34)\,(23) = (243)$.

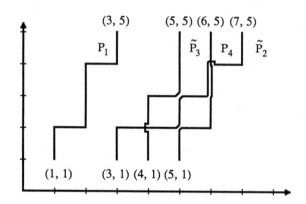

The involutions in §§4.2, 4.3 and 4.4 were not only sign-reversing, but also weight-preserving. Is there a version of Theorem 5.1 that involves weights, i.e, generating functions? Can we find a weight w such that $w(\sigma, (P_1, \dots, P_k)) = w(\varphi(\sigma, (P_1, \dots, P_k)))$? Since φ preserves all the edges in the lattice paths $(P_1, \dots, P_k)$, any weight which depends upon this set of edges will be preserved by φ. Let us describe such a weight. Given a path $P : (a, 1) \to (b, N)$, let $\mathrm{Hor}_y(P)$ be the multiset of the y-coordinates of the horizontal steps of P. Let

(5.3) $w(P) = \prod_{i \in Hor_y(P)} x_i$.

In this example

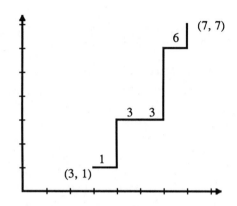

$Hor_y(P) = \{1, 3, 3, 6\}$ and $w(P) = x_1^1 x_3^2 x_6^1$.

Now extend w to S by $w(\sigma, (P_1, \ldots, P_k)) = w(P_1) w(P_2) \cdots w(P_k)$. By the preceding remarks, w is preserved by φ. We can now replace the binomial coefficients that appear in Theorem 5.1 by generating functions. Given $(a, 1)$ and (b, N), we saw that there are

$$\binom{N+b-a-1}{b-a}$$

lattice paths $P : (a, 1) \to (b, N)$. But we can also describe the generating function of all such paths:

(5.4) $\sum_P w(P)$.

Any path P is uniquely determined by its horizontal steps. Thus the terms in (5.4) will be monomials

$$x_1^{m_1} x_2^{m_2} \cdots x_N^{m_N}$$

such that $m_1 + m_2 + \ldots + m_N = b - a$. Note in particular that (5.4) depends only on $x_1, x_2, \ldots, x_N$ and $b - a$. Write

(5.5) $h_{b-a}(x_1, x_2,..., x_N) = \sum_P w(P).$

So Theorem 5.1 can be generalized as follows.

THEOREM 5.2   *Given integers* $0 < a_1 < a_2 < ... < a_k$ *and* $0 < b_1 < b_2 < ... < b_k$, *let* $M = (M_{ij})$ *be the* $k \times k$ *matrix with*

$$M_{ij} = h_{b_j - a_i}(x_1, x_2,..., x_N).$$

Then

$$\det(M) = \sum_{(P_1,...,P_k)} w(P_1) \cdots w(P_k),$$

where the sum is taken over all sequences $(P_1, ... , P_k)$ *of non-intersecting lattice paths* $P_i : (a_i, 1) \to (b_i, N)$.

The function $h_n(x_1, x_2, ... , x_N)$ is called the *complete* or *homogeneous* symmetric function of degree n; it has the generating function

(5.6) $\sum_{k=0}^{\infty} h_k(x_1, x_2,..., x_N) t^k = (1 - t x_1)^{-1}(1 - t x_2)^{-1}\cdots (1 - t x_N)^{-1}.$

There is a surprising connection between a special case of Theorem 5.2 and column strict tableaux. Put $a_i = i$ and $b_i = \lambda_{k+1-i}$, $1 \le i \le k$, for some partition λ with k parts, $\lambda_1 \ge \lambda_2 \ge ... \ge \lambda_k$.

PROPOSITION 5.3   *There is a weight-preserving bijection* φ *between non-intersecting paths* $(P_1, ... , P_k)$, $P_i : (i, 1) \to (\lambda_{k+1-i} + i, N)$ *and column strict tableaux of shape* λ *with entries from* [N].

Proof The bijection φ is easy to describe. Take as an example $k = 4$, $\lambda = (5, 3, 2, 2)$ and $N = 6$. Here is a set of 4 non-intersecting paths:

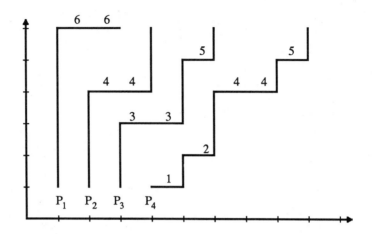

Each horizontal edge is labeled with its y-coordinate. Place these entries of path P_{k+1-i} into row i of the tableau T,

$$T = \begin{array}{|c|c|c|c|c|}
\hline
1 & 2 & 4 & 4 & 5 \\
\hline
3 & 3 & 5 \\
\cline{1-3}
4 & 4 \\
\cline{1-2}
6 & 6 \\
\cline{1-2}
\end{array}$$

Why is T column strict? Certainly it has shape λ and is weakly increasing across rows. Let T_{ij} be the entry in row i, column j of T. Then the paths P_{k-i} and P_{k-i+1} (which correspond to rows $i+1$ and i respectively in T) begin their jth horizontal edge at $x = k - i + j - 1$ and $x = k - i + j$ respectively. Since P_{k-i} is to the left of P_{k-i+1}, the jth horizontal edge of P_{k-i} must be strictly above the jth horizontal edge of P_{k-i+1}. Thus $T_{ij} < T_{i+1\,j}$.

The inverse of φ is also easy. Given a tableau T, the entries in row i of T determine the horizontal steps of P_{k-i+1}, and thus the entire path P_{k-i+1}. The paths again do not intersect because T is column strict.

The weight of a column strict tableau T, $w(T)$, is defined by

$$(5.7) \qquad w(T) = \prod_{\substack{\text{entries} \\ i \text{ of } T}} x_i \,.$$

In our example,

$$w(T) = x_1^1 x_2^1 x_3^2 x_4^4 x_5^2 x_6^2 = w(P_1, P_2, P_3, P_4).$$

This obviously makes φ weight-preserving.

Theorem 5.2 can now be applied to tableaux.

COROLLARY 5.4 *If λ is a partition of N with k parts,*

$$\sum_{\rho = (\rho_1, \ldots, \rho_N)} K_{\lambda\rho} x_1^{\rho_1} \cdots x_N^{\rho_N} = \det[h_{\lambda_i + j - i}(x_1, \ldots, x_N)],$$

where $1 \le i, j \le k$ and $K_{\lambda\rho}$ is the Kostka number of §3.5.

∎

In Corollary 5.2 of Chapter 3 we showed that $K_{\lambda\rho} = K_{\lambda\rho'}$, if ρ' is any reordering of ρ. This statement is obvious from Corollary 5.4, because each entry in the determinant is a symmetric polynomial in $x_1, x_2, \ldots, x_N$.

The symmetric polynomial in Corollary 5.4 is called the *Schur function* and is denoted $s_\lambda(x_1, \ldots, x_N)$.

Corollary 5.4 can now be used to prove a formula of Frobenius (1899) for the number of standard tableaux of shape λ.

PROPOSITION 5.5 *For any partition λ of N,*

$$d_\lambda = N! \prod_{t=1}^{k} [(\lambda_t + k - t)!]^{-1} \prod_{1 \le i < j \le k} (\lambda_i - i - \lambda_j + j),$$

where $\lambda_1 \ge \lambda_2 \ge \ldots \ge \lambda_k$ and $\lambda_1 + \lambda_2 + \ldots + \lambda_k = N$.

Note: For $\lambda = (4, 2, 2)$, Proposition 5.5 gives $d_{422} = 56$, which agrees with the result from the hook formula of §3.5. In fact, the Frobenius formula can be shown to be equivalent to the hook formula, Theorem 5.4 of Chapter 3. You are asked to do this in Exercise 29.

Proof Since $d_\lambda = K_{\lambda\rho}$, $\rho = 1^N$, we see that d_λ is the coefficient of $x_1 x_2 \cdots x_N$ in

(5.8) $\quad \det [h_{\lambda_i+j-i}(x_1,..., x_N)] = \sum_{\sigma \in S_k} \text{sgn}(\sigma) \prod_{i=1}^{k} h_{\lambda_i+\sigma(i)-i}(x_1,..., x_N).$

Such a monomial can be created by selecting a square-free monomial from each h_k, $k = \lambda_i - i + \sigma(i)$, with the product of these monomials equal to $x_1 x_2 \cdots x_N$. The number of ways of doing this is easily seen to be the multinomial coefficient

$$\binom{N}{\lambda_1+\sigma(1)-1,..., \lambda_k+\sigma(k)-k}.$$

So (5.8) implies

(5.9) $\quad d_\lambda = N! \det [[(\lambda_i - i + j)!]^{-1}], \ 1 \le i, j \le k.$

By factoring out entries in the last column, we get

(5.10) $\quad d_\lambda = N! \prod_{t=1}^{k} [(\lambda_t+k-t)!]^{-1} \det [(\lambda_i-i+j+1) \cdots (\lambda_i-i+k)],$

where $1 \le i, j \le k$. Let $P_{kj}(x)$ be the polynomial in x of degree $k-j$

(5.11) $\quad P_{kj}(x) = (x+j+1) \cdots (x+k)$

so that

(5.12) $\quad d_\lambda = N! \prod_{t=1}^{k} [(\lambda_t+k-t)!]^{-1} \det [P_{kj}(\lambda_i-i)], \ 1 \le i, j \le k.$

By column operations which do not change the value of the determinant,

(5.13) $\quad d_\lambda = N! \prod_{t=1}^{k} [(\lambda_t+k-t)!]^{-1} \det [(\lambda_i-i)^{k-j}], \ 1 \le i, j \le k.$

This determinant is Vandermonde's; the evaluation given in (2.1) gives Proposition 5.5.

The reader can try to evaluate $\det M$ for $x_1 = x_2 = ... = x_N = 1$ in Theorem 5.2. The result is the number of column strict tableaux of shape λ with entries in $[N]$.

§4.6 The Involution Principle

Thus far we have seen several examples of sign-reversing involutions on signed sets. Now suppose we have two signed sets, $A = A^+ \cup A^-$ and $B = B^+ \cup B^-$. We say that there is a *signed bijection* between A and B if there is a sign-reversing involution φ with no fixed-points on $A \cup B$, where $(A \cup B)^+ = A^+ \cup B^-$ and $(A \cup B)^- = A^- \cup B^+$. Notice that this implies that $\|A\| = \|B\|$. For example, any point in A^+ is identified with either a point in A^- (a cancellation) or a point in B^+. When A and B are not signed, i. e., $A^+ = A$ and $B^+ = B$, then φ is an ordinary bijection between A and B.

Garsia and Milne [Ga-M] discovered a general method of constructing signed bijections.

THEOREM 6.1 *Let* A *be a finite signed set,* $A = A^+ \cup A^-$, *with sign-reversing involutions* φ *and* ψ *whose fixed-point sets are* $F(\varphi)$ *and* $F(\psi)$ *respectively. Then there is a signed bijection* γ *between* $F(\varphi)$ *and* $F(\psi)$. *Furthermore,* γ *can be constructed using the following algorithm:*

ALGORITHM 21: *Involution Principle*

begin
    if $\varphi(x) = x$ **then**
        $y \leftarrow x$
        repeat
            $z \leftarrow \psi(y)$
            $y \leftarrow \varphi(z)$
        until $\varphi(z) = z$ **or** $\psi(y) = y$
        if $\varphi(z) = z$ **then**
            $\gamma(x) \leftarrow z$
        else
            $\gamma(x) \leftarrow y$
    else if $\psi(x) = x$ **then**
        $y \leftarrow x$
        repeat
            $z \leftarrow \varphi(y)$
            $y \leftarrow \psi(z)$
        until $\psi(z) = z$ **or** $\varphi(y) = y$
        if $\psi(z) = z$ **then**
            $\gamma(x) \leftarrow z$

else

$$\gamma(x) \leftarrow y$$

else

$\{x$ is not a fixed point of φ or $\psi\}$

end.

Proof We construct a graph G whose vertices are elements of A. The edges of G are labeled φ or ψ, with the φ-edges given by $(x, \varphi(x))$ for $x \in A - F(\varphi)$ and the ψ-edges by $(x, \psi(x))$ for $x \in A - F(\psi)$. For example, if the involution φ is given by this picture:

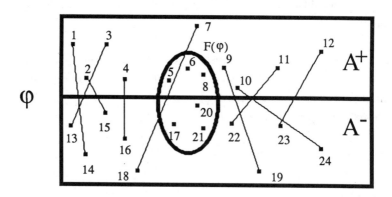

and the involution ψ by this picture:

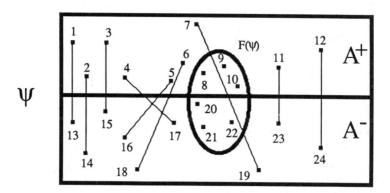

then the graph G looks like this:

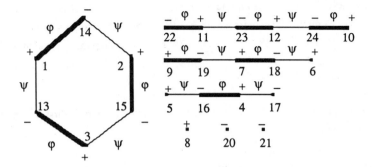

Clearly the degree of any vertex of G is at most 2, so the connected components of G consist of isolated points, chains or cycles. Also, elements of $F(\varphi)$ and $F(\psi)$ will be isolated points or endpoints of chains. If $x \in A$ is isolated, then $x \in F(\varphi) \cap F(\psi)$ so Algorithm 21 will give $\gamma(x) = x$.

Suppose $x \in F(\varphi)$ is the end of a chain. The edges of this chain must alternate labels φ and ψ. If the last edge in the chain is labeled ψ, the other endpoint y of the chain must be in $F(\varphi)$. In this case there will be an odd number of edges in the chain, so x and y have opposite signs. In the example above, $x = 5$ and $y = 17$ are connected by such a chain.

If the last edge in the chain is labeled φ, the other endpoint y of the chain must be in $F(\psi)$. There will be an even number of edges in the chain, so x and y have the same signs. In the example above, $x = 6$ and $y = 9$ is such a chain.

Note that Algorithm 21 merely constructs values along the chain until the other end is reached.

Frequently, the involution principle gives an explicit bijection between two apparently unrelated sets. In practice, the sets $F(\varphi)$ and $F(\psi)$ may be very small subsets of A. The number of iterations in Algorithm 21 can be very large, and depends on the element chosen.

Garsia and Milne used the involution principle to give the first proof using a bijection of the *Rogers-Ramanujan identities*.

THEOREM 6.2 *The number of partitions of* n *whose parts are congruent to* 1 *or* 4 mod 5 *is the same as the number of partitions of* n *into distinct parts whose consecutive parts differ by at least two.*

Garsia and Milne discovered a large signed set A which contained the two sets of partitions in question, and they found two involutions whose fixed point sets were these two sets. Thus, they had found a bijection between the two sets.

As an example of Theorem 6.2, if $n = 12$, there are 9 partitions in each set: $11\,1$, $9\,1^3$, 6^2, $6\,4\,1^2$, $6\,1^6$, 4^3, $4^2\,1^4$, $4\,1^8$ and 1^{12} with parts congruent to 1 or 4 mod 5; and 12, $11\,1$, $10\,2$, $9\,3$, $8\,4$, $8\,3\,1$, $7\,5$, $7\,4\,1$ and $6\,4\,2$ whose parts differ by at least two.

We will now give three examples of the involution principle. First we prove Euler's theorem.

THEOREM 6.3 *The number of partitions of* n *into odd parts equals the number of partitions of* n *into distinct parts.*

Proof Let P_k be the set of all partitions of k and let ED_{n-k} be the set of partitions of $n-k$ into even, distinct parts. Let

$$A = \bigcup_{k=0}^{n} P_k \times ED_{n-k}$$

and define the sign of an element $x = (p_1, p_2) \in A$ by

$$sgn(x) = (-1)^{\text{number of parts of } p_2}.$$

For the involution φ, take the smallest even part e of p_1 or p_2 and move e from p_1 to p_2 if e is not in p_2. Otherwise move e from p_2 to p_1. Clearly φ changes the number of parts of p_2 by one, so φ is sign-reversing. Also, $F(\varphi) = \{(p_1, \varnothing) : p_1 \text{ has only odd parts}\} \subset A^+$. As an example, let $n = 26$ and $k = 12$:

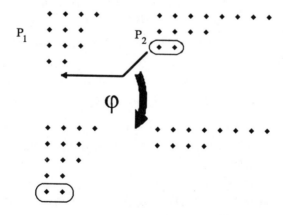

Now we need the involution ψ whose fixed-point set $F(\psi)$ is $\{(p_1, \varnothing) : p_1$ has distinct parts$\}$. Given (p_1, p_2), let (i, i) be the smallest repeated part in p_1 and let $2j$ be the smallest part of p_2. If $i < j$ or $p_2 = \varnothing$, move (i, i) from p_1 to p_2 by creating a part of size $2i$; if $i \geq j$, move $2j$ from p_2 to p_1 by creating two parts, (j, j), in p_1. In the example below, $n = 26$ and $k = 12$.

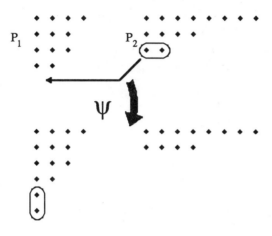

Again ψ changes the number of parts of p_2 by one. The fixed point set is what we wanted. This completes the proof.

∎

The involution principle can be extended so that the involutions φ and ψ act on different signed sets A and B respectively. If there is a signed bijection ζ between A and B, then there is a signed bijection between $F(\varphi)$ and $F(\psi)$. In Algorithm 21, replace φ and ψ with sign-reversing versions of ζ, $\varphi \circ \zeta$ and $\psi \circ \zeta$. The proof of Theorem 6.1 is much the same, and the picture below gives an example. You are asked in Exercise 31 to do a related problem.

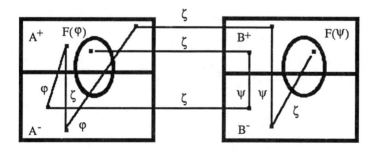

Remmel [Re] has used this version of the involution principle to give several bijections. Here are two. The first is a theorem due to Schur.

THEOREM 6.3 *The number of partitions of* n *into parts congruent to* 1 *or* 5 mod 6 *equals the number of partitions of* n *into distinct parts congruent to* 1 *or* 2 mod 3.

Proof The idea of the proof is to define two signed sets, A and B: A related to the mod 6 condition and B to the mod 3 condition. Let us begin with A. Since $F(\varphi) \subset A^+$, the involution φ should move any partition in A with parts 2, 3, 4, 6, 8, ... For any partition λ of n, let $S_A(\lambda)$ be the set of these "illegal" parts of λ. For example, if $\lambda = 9\ 8^2\ 7^3\ 6\ 2\ 1^2$, then $S = \{9, 8, 6, 2\}$.

Now define $A = \{(\lambda, S) : \lambda$ partitions n, $S \subset S_A(\lambda)\}$ and put $\text{sgn}(\lambda, S) = (-1)^{|S|}$. The involution φ on A merely changes S. Suppose $(\lambda, S) \in A$. If $S_A(\lambda)$ is non-empty, insert (delete) its largest element into (from) S to obtain $\tilde{S}$; $\varphi(\lambda, S) = (\lambda, \tilde{S})$. For example,

$$\varphi(\ 9\ 8^2\ 7^3\ 6\ 2\ 1^2, \{9, 8, 2\}) = (\ 9\ 8^2\ 7^3\ 6\ 2\ 1^2, \{8, 2\}).$$

The fixed points of φ are those (λ, S) such that $S_A(\lambda)$ is empty, that is, λ has no "illegal" parts.

The set B is defined in a similar way. The "illegal" parts are somewhat more complicated. There are two kinds of illegal parts: multiples of three, i. e., 3, 6, 9, ... ; and <u>pairs</u> of equal parts congruent to 1 or 2 mod 3, i. e., $1^2, 2^2, 4^2, 5^2, \dots$. For any partition λ of n, let

$$I_1(\lambda) = \{k : k \text{ is a part of } \lambda \text{ and } k \equiv 0 \mod 3\} \text{ and}$$
$$I_2(\lambda) = \{j^2 : j \text{ is a repeated part of } \lambda, \ j \equiv 1 \text{ or } 2 \mod 3\}.$$

Let $S_B(\lambda) = I_1(\lambda) \cup I_2(\lambda)$. Put $B = \{(\lambda, S) : \lambda$ partitions n, $S \subset S_B(\lambda)\}$ and put $\text{sgn}(\lambda, S) = (-1)^{|S|}$. Again, the involution ψ either inserts or deletes an element from S. To see which element, we need a weight w on $S_B(\lambda)$. If $k \in I_1(\lambda)$, let $w(k) = k$; if $j^2 \in I_2(\lambda)$, let $w(j^2) = j + j$. Given $(\lambda, S) \in B$, if $S_B(\lambda)$ is non-empty, either delete or insert the element of $S_B(\lambda)$ with largest weight from S. For example, if $\lambda = 6\ 5\ 4^3\ 3\ 1$ and $S = \{6, 4^2\}$, then $\psi(\lambda, S) = (6\ 5\ 4^3\ 3\ 1, \{6\})$, since $S_B(\lambda) = \{6, 4^2, 3\}$ and $w(4^2) = 8$ is the largest weight in $S_B(\lambda)$. The fixed points of ψ are those (λ, S) such that $S_B(\lambda)$ is empty.

It remains to construct the signed bijection ζ between A and B. But all that is necessary is to identify the elements $x \in S_A(\lambda)$ with elements $\tilde{x} \in S_B(\lambda)$ in such a way that $x = w(\tilde{x})$. This bijection is

$$6i+2 \leftrightarrow (3i+1)^2;$$
$$6i+4 \leftrightarrow (3i+2)^2;$$
$$6i \leftrightarrow 6i;$$
$$6i+3 \leftrightarrow 6i+3.$$

So for $(\lambda, S) \in A$, let $\zeta(\lambda, S) = ((\lambda - S) \cup \tilde{S}, \tilde{S})$ where $\tilde{S}$ is constructed from S by the bijection above. Then ζ identifies elements of A with elements of B. Since $|S| = |\tilde{S}|$, the sign is preserved and since the weights are maintained, n is unchanged. For example,

$$\zeta(9\ 8^2\ 7^3\ 6\ 2\ 1^2, \{9, 8, 2\}) = (9\ 8\ 7^3\ 6\ 4^2\ 1^4, \{9, 4^2, 1^2\}).$$

For the last example we take Exercise 16 of Chapter 1: the number of partitions of n whose even parts must be distinct is equal to the number of partitions of n such that no part is repeated more than three times. Using the previous example as a model, we need only identify the "illegal" parts. For the set A, the illegal parts are pairs of even parts: $\{2^2, 4^2, 6^2, \dots\}$; for the set B, they are quadruples: $\{1^4, 2^4, 3^4, \dots\}$. The bijection between these two sets is obvious. Thus, the involutions φ and ψ and signed bijection ζ can be defined as before.

Notes

As the reader can see, there has been much recent interest in involutions. These techniques have become particularly popular among a group of French and French Canadian mathematicians, including Foata, Joyal, G. Labelle, J. Labelle, Leroux and Viennot. Other mathematicians who have developed and used these techniques are Garsia, Gessel, Milne, Wilf and Zeilberger, to name just a few.

Andre's example appears in Feller [Fe].

Equation (5.8) is called the *Jacobi-Trudi identity*. Schur functions have many important applications in and outside mathematics and are closely related to the character theory of the symmetric group. Moreover, sign-reversing involutions seem to arise frequently and naturally in this theory.

Remmel [Re] has given several other applications of the involution principle. New applications of the principle are being found with increasing frequency. Many of these are too involved to be given here.

Exercises

1.[2]   Let A be the Boolean algebra B_n with $sgn(\alpha) = (-1)^{|\alpha|}$. Find a sign-reversing involution φ on A such that $|\alpha|$ and $|\varphi(\alpha)|$ differ by exactly one for all $\alpha \in A$. Use the involution given in the introduction to this chapter which proved the principle of inclusion-exclusion. What famous identity involving binomial coefficients have you proved?

2.[2]   Let A be the k-element subsets of [n], with n even. For each $\alpha \in A$, define $val(\alpha)$ to be

$$val(\alpha) = \sum_{i \in \alpha} i .$$

Let $A^+ = \{\alpha \in A : val(\alpha)$ is even$\}$ and $A^- = \{\alpha \in A : val(\alpha)$ is odd$\}$. Construct a sign-reversing involution φ on A which will prove

$$|A^+| - |A^-| = \begin{cases} 0 & \text{if } k \text{ is odd} \\ (-1)^{k/2} \binom{n/2}{k/2} & \text{if } k \text{ is even.} \end{cases}$$

In each of the next five exercises, give a generating function proof of the identity and then prove it combinatorially with a sign-reversing involution. In each case all the parameters may be considered positive integers. Recall from Chapters 1 and 3 the combinatorial interpretations of binomial coefficients and Stirling numbers.

3.[2]

$$\sum_{k=p}^{n} \binom{n}{k} \binom{k}{p} (-1)^k = \delta_{np} (-1)^p .$$

4.[3]

$$\sum_{j=0}^{m} \binom{n}{j} \binom{n}{m-j} (-1)^j = \begin{cases} 0 & \text{if } m \text{ is odd,} \\ \binom{n}{k} (-1)^k & \text{if } m = 2k. \end{cases}$$

5.[3]

$$\sum_{k=0}^{m} \binom{n+k-1}{k} \binom{n}{m-k} (-1)^{m-k} = \delta_{m0}.$$

6.[3]

$$\sum_{k=j}^{n} S(n, k)\, s(k, j) = \delta_{nj}.$$

7.[3]

$$\sum_{k=j}^{n} s(n, k)\, S(k, j) = \delta_{nj}.$$

8.[3]   Give a combinatorial proof using a weight-preserving, sign-reversing involution.

$$x^{n} = \sum_{k=0}^{n} S(n, k)\, (x)^{(k)}\, (-1)^{n-k}.$$

9.[3]   Give a combinatorial proof using a weight-preserving, sign-reversing involution.

$$(x)_{n} = \sum_{k=0}^{n} s(n, k)\, x^{k}.$$

10.[2]   Give a sign-reversing involution which proves the more general version of the principle of inclusion-exclusion:

$$N_{=}(S) = \sum_{S \subset T \subset P} (-1)^{|T|-|S|}\, N_{\supseteq}(T).$$

11.[2]   Let P be the set of all partitions and PD be the set of all partitions with distinct parts. Let $\|p\|$ denote the number that $p \in P$ partitions. Then we know

$$\sum_{p \in P} x^{\|p\|} = \frac{1}{(1-x)(1-x^{2}) \cdots}.$$

As in the Euler pentagonal number theorem, PD is a signed set and

$$\sum_{p \,\in\, PD} \text{sgn}(p)\, x^{\|p\|} \;=\; (1-x)(1-x^2) \cdots \,.$$

Prove that

$$\frac{(1-x)(1-x^2) \cdots}{(1-x)(1-x^2) \cdots} \;=\; 1,$$

by finding an appropriate involution on the set of ordered pairs (p_1, p_2), $p_1 \in P$ and $p_2 \in PD$.

12.[2]   It is also clear that

$$\frac{(1-x^2)(1-x^4) \cdots}{(1-x)(1-x^2) \cdots} \;=\; \frac{1}{(1-x)(1-x^3) \cdots}\,.$$

Define a signed set and an involution φ which proves this identity. You should consider ordered pairs of partitions, as in Exercise 11.

13.[2]   Define a signed set and an involution φ which proves this identity:

$$\frac{(1-x)(1-x^2) \cdots (1-x^n)}{(1-x)(1-x^2) \cdots} \;=\; \frac{1}{(1-x^{n+1})(1-x^{n+2}) \cdots}\,.$$

As in Exercises 11 and 12, consider ordered pairs of partitions.

14.[3]   Write down eight different versions of the identity in Exercise 13. For example, here is another:

$$\frac{(1-x^{n+1})(1-x^{n+2}) \cdots}{(1-x)(1-x^2) \cdots} \;=\; \frac{1}{(1-x)(1-x^2) \cdots (1-x^n)}\,.$$

Interpret each of these as an identity involving partitions and give a combinatorial proof of each. Some will require sign-reversing involutions.

15.[3]   Define a signed set and an involution φ which proves:

$$\frac{1}{(1-x)(1-x^2) \cdots} \; \frac{1}{(1+x)(1+x^2) \cdots} \;=\; \frac{1}{(1-x^2)(1-x^4) \cdots}\,.$$

As in Exercises 11 and 12, consider ordered pairs of partitions.

16.[3]   Repeat Exercise 14 for the identity in Exercise 15.

17.[2]   Prove Proposition 2.1.

18.[1]   Let $\bar{a}$ be the score vector of a tournament T. Show that if exactly one three-cycle of T is reversed, $\bar{a}$ remains unchanged.

19.[2]   (Gessel [Ge]) Let G be the bipartite graph whose vertices are the non-transitive tournaments on [n], and edges $T - \tilde{T}$ if $\tilde{T}$ can be obtained from T by reversing exactly one three-cycle. Show that any connected component of G is regular. (An important theorem from graph theory, the matching theorem, then implies that G has a complete match.)

20.[3]   The purpose of this exercise is to show that the inversion poset $\mathcal{l}_n$ is a lattice. For π and σ in $\mathcal{l}_n$, we must define their join and meet. Let T_π be the transitive tournament with ranking π. Color the edges of T_π blue and red: $i \rightarrow j$, $i < j$ is blue; $i \rightarrow j$, $i > j$ is red (the *upsets*).

(a) Show that σ covers π if and only if T_σ and T_π are identical, except for a blue directed edge of T_π which is red and reversed in T_σ.

(b) For a given π and σ, show that any transitive tournament T whose red edges contain the red edges of T_π and T_σ corresponds to a permutation which lies above π and σ in $\mathcal{l}_n$.

(c) Show that there is a unique transitive tournament T satisfying (b) whose red set is minimal. This tournament corresponds to the join of π and σ.

(d) How do you define the meet of π and σ?

21.[3]   (Zeilberger [Z1]) Let S_n denote the permutations of n and for $\pi \in S_n$, put $w(\pi) = a_{1\pi(1)} \cdots a_{n\pi(n)}$, so that

$$\det(A) = \sum_{\pi \in S_n} \text{sgn}(\pi)\, w(\pi).$$

(a) Define a weight-preserving, sign-reversing involution on S_n which proves $\det A = 0$ if $a_{1i} = a_{2i}$ for $1 \leq i \leq n$.

(b) Give a combinatorial proof of $\det (AB) = (\det A)(\det B)$.

(c) Define $A^* = (\gamma_{ij})$ with $\gamma_{ij} = (-1)^{i+j} \det (A_{ji})$, where A_{ji} is the ji-th minor. Give a combinatorial proof that $A^* A = (\det A) I$.

(d) Give a combinatorial proof of the expansion formula for $\det A$ along the jth row of A.

22.[2]   Prove Cayley's Theorem (Theorem 2.1 of Chapter 3) using Theorem 4.1.

23.[1]   Let $h_k(x_1, \ldots , x_N)$ be the complete homogeneous symmetric function in $x_1, \ldots , x_N$ of degree k. From Section 5 we see that $h_k(1, 1, \ldots , 1) =$

$$\binom{N+k}{k}.$$

What is $h_k(q^1, q^2, \ldots , q^N)$?

24.[2]   Use Exercise 23 and (5.8) to find a determinantal expression for

$$\sum_P q^{\|P\|},$$

where the sum is over all column strict tableaux P of shape λ and $\|P\|$ is the sum of entries in P. Can you evaluate your determinant?

25.[1]   Let μ be a partition of n, $\mu_1 \geq \mu_2 \geq \ldots \geq \mu_k$. Define the homogeneous symmetric function

$$h_\mu(x_1, \ldots, x_N) = h_{\mu_1} h_{\mu_2} \cdots .$$

Interpret $h_\mu(x_1, \ldots , x_N)$ as a generating function for a class of multiset permutations.

26.[2]   Use the Schensted correspondence (Chapter 3) to conclude

$$h_\mu = \sum_\lambda K_{\lambda\mu} s_\lambda ,$$

where s_λ is as defined in §4.5.

27.[3]   Define the elementary symmetric functions, $e_j(x_1, \ldots, x_N)$:

$$e_j(x_1, \ldots, x_N) = \sum_{\substack{\{i_1, i_2, \ldots, i_j\} \\ \text{distinct} \\ \text{in } [N]}} x_{i_1} x_{i_2} \cdots x_{i_j} \;,$$

and, as in Exercise 25, define $e_\mu(x_1, \ldots, x_N)$:

$$e_\mu(x_1, \ldots, x_N) = e_{\mu_1} e_{\mu_2} \cdots \;.$$

Interpret $e_\mu(x_1, \ldots, x_N)$ as an appropriate generating function and use the Schensted correspondence to prove

$$e_\mu = \sum_\lambda K_{\lambda\mu} s_{\lambda'} \;.$$

28.[3]   This exercise shows that

$$\Delta(x_1, \ldots, x_N) \, e_\mu = \sum_\lambda K_{\lambda\mu} \det [\, x_i^{\lambda'_j + N - j} \,],$$

where $\Delta(x_1, \ldots, x_N)$ is Vandermonde's determinant. Write the left-hand side as

$$\sum_{(\pi, A_1, \ldots, A_k)} w(\pi) w(A_1) \cdots w(A_k) \operatorname{sgn}(\pi),$$

where $\pi \in S_N$ (see Exercise 21), $A_i \subset [N]$, $|A_i| = \mu_i$,

$$w(\pi) = x_{\pi(1)}^{N-1} \cdots x_{\pi(N)}^0$$

and

$$w(A_i) = \prod_{j \in A_i} x_j \;.$$

Find a sign-reversing involution φ on this set such that $F(\varphi)$ consists of those $(\pi, A_1, \ldots, A_k)$ such that for all $1 \le j \le k$, $w(\pi) w(A_1) \cdots w(A_j)$ has distinct exponents. Show that $F(\varphi)$ is exactly what is counted by the right-hand side.

29.[3]   Prove the hook formula (Theorem 5.4 of Chapter 3) using the Frobenius formula (Proposition 5.5).

30.[1]   Show that the existence of sign-reversing involutions φ and ψ immediately implies $|F(\varphi)| = |F(\psi)|$, without the explicit bijection given by the involution principle.

31.[2]   Suppose A, B and C are three signed sets and suppose φ and ψ are signed bijections between A and B and between B and C, respectively.

(a) Prove that there is a signed bijection between A and C, which might be considered the "composition" of φ and ψ.

(b) The signed bijections φ and ψ might be degenerate in some sense. For example, φ might be a "pure" bijection, i. e., $\varphi(A^+) = B^+$ and $\varphi(A^-) = B^-$. Or φ might be a sign-reversing involution on A or B: $\varphi(B^+) \subset A^+$ and $\varphi(B^-) \subset A^-$. Determine under what conditions the composition described in (a) requires the involution principle.

32.[4C]   Program the involution principle and apply your program to Euler's theorem. Run your program for various values of n. How does the bijection compare with the bijection given in Chapter 3? Guess and prove the theorem.

33.[4C]   Suppose A is a signed set and X, Y $\subset$ A, with $|X| = |Y| = \|A\|$ and $X \cap Y = \emptyset$. Write a program to construct two random involutions φ and ψ such that $F(\varphi) = X$ and $F(\psi) = Y$. Investigate

(a) the average length of a path from an element in X to an element in Y; and

(b) the average number of cycles in the graph of φ and ψ.

What conjectures can you make and what theorems can you prove?

34.[4]   This problem is due to Blas and Sagan [Bl-Sa] and Zeilberger [Z4]. Let G be a simple graph with vertex set V(G) and edge set E(G). A *proper coloring* of G with k *colors* is a function from V(G) to the colors such that no two adjacent vertices are given the same color. The *chromatic polynomial* of G is the function $P_G(x)$, the number of ways of properly coloring V(G) with x colors. For example, the chromatic polynomial of K_n is $(x)_n$. (It is easy to see that $p_G(x)$ is a polynomial.) Whitney's broken circuit theorem [Whn] gives a combinatorial interpretation of the coefficients of the chromatic polynomial in terms of broken circuits. Suppose the edges of G are ordered in some way. A *broken circuit* of G is a cycle with the largest edge of the cycle removed. Use the involution principle to prove the broken circuit theorem: if $|V(G)| = n$, the coefficient of k^{n-i} in $p_G(k)$ is the number of edge subsets of size i which do not contain a broken circuit. Hint: Let $A = \{(S, f) : S \subset E(G)$ and f is any coloring of $V(G)$ which is constant on the connected components of the graph generated by $S\}$. Make A into a signed set by defining $\text{sgn}(S, f) = (-1)^{|S|}$. Find an involution φ with $F(\varphi) = \{(\varnothing, f) \subset A : f$ is a proper coloring of $G\}$ and another involution ψ with $F(\psi) = \{(S, f) \subset A : S$ contains no broken circuits$\}$.

Bibliography

Undergraduate Texts in Combinatorics

[Bo] K. Bogart, *Introductory Combinatorics*, Pitman, Pitman, Massachusetts, 1983.

[Br] R. Brualdi, *Introductory Combinatorics*, North Holland, New York, 1978.

[Li] C. Liu, *Introduction to Combinatorial Mathematics*, McGraw-Hill, New York, 1968.

[Ro] F. Roberts, *Applied Combinatorics*, Prentice-Hall, Englewood Cliffs, New Jersey, 1984.

[Tu] A. Tucker, *Applied Combinatorics*, Second Edition, Wiley, New York, 1984.

Graduate Texts in Combinatorics

[Ai] M. Aigner, *Combinatorial Theory*, Springer, New York, 1979.

[Be] C. Berge, *Principles of Combinatorics*, Academic Press, New York, 1971.

[Co] L. Comtet, *Advanced Combinatorics*, Reidel, Dordrect, Boston, 1974.

[G-J] I. Goulden and D. Jackson, *Combinatorial Enumeration*, Wiley-Interscience, New York, 1983.

[Wi] S. G. Williamson, *Combinatorics for Computer Science*, Computer Science Press, Rockville, Maryland, 1985.

Texts on Combinatorial Algorithms

[E] S. Even, *Combinatorial Algorithms*, Macmillan, New York, 1973.

[Kn] D. Knuth, *The Art of Computer Programming*, Vol. 3, Addison-Wesley, Reading, Massachusetts, 1973.

[N-W] A. Nijenhuis and H. Wilf, *Combinatorial Algorithms*, Academic Press, New York, 1978.

[R-N-D] E. Reingold, J. Nievergelt, and N. Deo, *Combinatorial Algorithms*, Prentice-Hall, Englewood Cliffs, New Jersey, 1977.

Other References

[Ai] M. Aigner, Lexicographic matching in Boolean algebras, *J. Comb. Th. B* **14** (1973), 187-194.

[An] G. Andrews, *The Theory of Partitions*, Addison-Wesley, Reading, Massachusetts, 1976.

[B-Z] E. Bender and D. Zeilberger, Some asymptotic bijections, *J. Comb. Th. A* **38** (1985), 96-98.

[Bl-Sa] A. Blas and B. Sagan, Bijective proofs of two broken circuit theorems, *J. Graph Th.*, to appear.

[Bre-Z] D. Bressoud and D. Zeilberger, Bijecting Euler's partitions-recurrence, *Amer. Math. Monthly*
  92 (1985), 54-55.

[Ca] R. Canfield, On a problem of Rota, *Adv. in Math.* **29** (1978), 1-10.

[Ch] S. Chaiken, A combinatorial proof of the all minors matrix tree theorem, *SIAM J. Alg. Disc.
  Meth.* **3** (1982), 319-329.

[deB-T-K] N. deBruijn, C. van E. Tengbergen, and D. Kruyswijk, On the set of divisors of a number,
  Nieuw Arch. Wisk. (2) **23** (1952), 191-193.

[Eh] G. Ehrlich, Loopless algorithms for generating permutations, combinations, and other
  combinatorial configurations, *J. Assoc. Comput. Mach.* **20** (1973), 500-513.

[Fe] W. Feller, *An Introduction to Probability Theory and its Applications*, Vol. 1, Wiley, New
  York, 1957.

[Fo] D. Foata, On the Netto inversion number of a sequence, *Proc. Amer. Math. Soc.* **19** (1968),
  236-240.

[Fo-Schü] D. Foata and M. Schützenberger, Major index and inversion number of permutations,
  Math. Nachr. **83** (1978), 143-159.

[Fr] P. Frankl, A new short proof for the Kruskal-Katona Theorem, *Disc. Math.* **48** (1984), 327-329.

[Frz-Z] D. Franzblau and D. Zeilberger, A bijective proof of the hook-length formula, *J. Algorithms*
  3 (1982), 317-342.

[Ga-M] A. Garsia and S. Milne, Method for constructing bijections for classical partition identities,
  Proc. Natl. Acad. Sci. USA **78** (1981), 2026-2028.

[Ge] I. Gessel, Tournaments and Vandermonde's determinant, *J. Graph Theory* **3** (1979), 305-307.

[Ge-V] I. Gessel and G. Viennot, Binomial determinants, paths, and hook length formulae, *Adv. in
  Math.*, to appear.

[Gra] F. Gray, Pulse Code Communication, US Patent 2632058, March 17, 1953.

[Gre-K1] C. Greene and D. Kleitman, Proof techniques in the theory of finite sets, in *Studies in
  Combinatorics*, ed. by G.-C. Rota, Mathematical Association of America, 1978, 22-79.

[Gre-K2] C. Greene and D. Kleitman, Strong versions of Sperner's theorem, *J. Comb. Th.* **20**
  (1976), 80-88.

[Ja-K] G. James and A. Kerber, *The Representation Theory of the Symmetric Group*,
  Addison-Wesley, Reading, Massachusetts, 1981.

[Joh] S. Johnson, Generation of permutations by adjacent transpositions, *Math. Comp.* **17** (1963),
  282-285.

[Jo-Wh-Wi] J. Joichi, D. White and S. Williamson, Combinatorial Gray codes, *SIAM J. Comp.* **9**
  (1980), 130-141.

[Kn1] D. Knuth, Permutations, matrices and generalized Young tableaux, *Pac. J. Math.* **34** (1970),
  709-727.

[La] G. Labelle, Une nouvelle demonstration combinatoire des formules d'inversion de Lagrange, *Adv.
  in Math.* **42** (1981), 217-247.

[Li-O] J. Littlewood and C. Offord, On the number of real roots of a random algebraic equation (III),
  Mat. USSR Sb. **12** (1943), 277-285.

[Lot] M. Lothaire, *Combinatorics on Words*, Addison-Wesley, Reading, Massachusetts, 1983.

[Lov] L. Lovasz, *Combinatorial Problems and Exercises*, North-Holland, New York, 1979.

[Md] I. Macdonald, *Symmetric functions and Hall polynomials*, Oxford University Press, 1979.

[MM] P. MacMahon, *Combinatory Analysis*, Vol. 1 (1917) and Vol. 2 (1918), reprinted by Chelsea,
  New York, third edition, 1984.

[Mi] S. Milne, Restricted growth functions and incidence relations of the lattice of partitions of an
  n-set, *Adv. in Math.* **26** (1977), 290-305.

[Mo] J. Moon, *Counting Labelled Trees*, Canadian Math. Monographs, No. 1, 1970.

[Po] M. Pouzet, Application d'une propriété combinatoire des parties d'un ensemble aux groupes et aux relations, *Math Z.* **150** (1976), 117-134.

[Ra] G. Raney, Functional composition patterns and power series reversion, *Trans. Amer. Math. Soc.* **94** (1960), 441-451.

[Re] J. Remmel, Bijective proofs of classical partition identities, *J. Comb. Th. A* **33** (1982), 273-286.

[Ri] J. Riordan, *An Introduction to Combinatorial Analysis*, Princeton University Press, 1980.

[Sch] C. Schensted, Longest increasing and decreasing subsequences, *Canad. J. Math.* **13** (1961), 179-191.

[Sc] M. P. Schützenberger, La correspondance de Robinson, *Combinatoire et Représentation du Groupe Symétrique*, Strasbourg, 1976 (D. Foata, Ed.), 59-113, Lecture Notes in Mathematics, No. 579, Springer-Verlag, Berlin, 1977.

[Sta1] R. Stanley, Weyl groups, the hard Lefsheftz theorem, and the Sperner property, *SIAM J. Alg. Disc. Meth.* **1** (1980), 168-184.

[Sta2] R. Stanley, On the number of reduced decompositions of elements of Coxeter groups, *Europ. J. Combinatorics* **5** (1984), 359-372.

[Sta3] R. Stanley, Theory and applications of plane partitions: part 1, *Stud. Appl. Math* **1** (1971), 167-188.

[Sta4] R. Stanley, Theory and applications of plane partitions: part 2, *Stud. Appl. Math* **1** (1971), 259-279.

[Str] H. Straubing, A combinatorial proof of the Cayley-Hamilton theorem, *Disc. Math.* **43** (1983), 273-279.

[T] H. Trotter, Algorithm 115: Perm, *Comm. ACM* **5** (1962), 434-435.

[Wh-Wi] D. White and S. G. Williamson, Recursive matching algorithms and linear orders on the subset lattice, *J. Comb. Th. A* **23** (1977), 117-127.

[Whn] H. Whitney, A logical expansion in mathematics, *Bull. Amer. Math. Soc.* **38** (1932), 572-579.

[Z1] D. Zeilberger, A combinatorial approach to matrix algebra, *Disc. Math.* **56** (1985), 61-72.

[Z2] D. Zeilberger, Garsia's and Milne's bijective proof of the inclusion-exclusion principle, *Disc. Math.* **51** (1984), 109-110.

[Z3] D. Zeilberger, A truly refined bijection among trees, to appear.

[Z4] D. Zeilberger, personal communication.

Appendix

In this appendix we give Pascal procedures for each of the algorithms in the text. We have grouped the algorithms roughly according to topic. For instance, the Johnson-Trotter algorithm (Algorithm 1) together with its Rank and Unrank algorithms (Algorithms 2 and 3) all appear in the section called Permutations.

This appendix contains only the subroutines associated with the algorithms in the text. Input-output programs and drivers have been omitted.

In order to simplify the parameter passing to the algorithms, some structured data types (e. g., Pascal records) are used. For instance, the data type *permutation* contains, in addition to the permutation itself, the inverse and the direction vector and the activity set. Furthermore, certain parameters have been determined to be global, e. g., N for the listing, ranking and unranking of permutations of [N].

As much as possible, standard Pascal has been used. When non-standard constructs appear, we clearly identify them. The procedures were developed on Macintosh Pascal©.

The form of the listing algorithms has been changed somewhat from what appears in the text to isolate input/output as much as possible. Instead of a single program which lists all relevant objects, two subroutines are used: GetFirst and GetNext. The parameters used by these two procedures are the object and a boolean variable which is **true** if no more objects are in the list and **false** otherwise. The listing programs are all then virtually the same, except for changes in data types.

A.1 Permutations

Note the structured data type for permutations. The permutation length is a global variable.

```
const
MaxLength = 12;

type
Vector = array[0..MaxLength] of integer;
IndexSet = set of 0..MaxLength;
Permutation = record
 Inverse, Value, Direction : Vector;
 Activity : IndexSet;                    {set of active indices}
 end; {Permutation}

var
N : integer;                            {length of permutation}
```

```
procedure UnrankPerm (Rnk : integer; var Pi : Permutation);
var
j, Dir, PrevRank, Remainder, Count, PrevN : integer;

begin
with Pi do
 begin
 for j := 1 to N do                    {initialize permutation}
  Value[j] := 0;
 PrevRank := Rnk;
 for PrevN := N downto 1 do
  begin
  Remainder := PrevRank mod PrevN;      {amount moved up or down}
  PrevRank := PrevRank div PrevN;       {rank of PrevN-1}
  if (PrevRank mod 2 = 1) then          {even means PrevN moving left; odd means right}
   begin
   j := 0;                             {initialize at left}
   Dir := 1;                           {moving right}
   end {if then}
  else
   begin
   j := N + 1;                         {initialize at right}
   Dir := -1;                          {moving left}
   end; {else}
  Count := 0;
  repeat
  j := j + Dir;                        {advance left or right one position}
  if (Value[j] = 0) then
   Count := Count + 1;                 {advance count for each index not assigned}
  until (Count = Remainder + 1);       {quit when count reaches amount to be moved}
  Value[j] := PrevN;
  end; {for}
 end; {with}
end; {UnrankPerm}

procedure RankPerm (Pi : Permutation; var Rnk : integer);
var
i, Moves, Remainder : integer;

function MoveCount (p : integer) : integer;   {returns number of numbers <p and left of p in Pi}
var
j, Count : integer;
begin
Count := 0;
with Pi do
 for j := 1 to Inverse[p] do          {Look at values left of p in Pi}
  if (Value[j] < p) then              {increase Count if they are <p}
   Count := Count + 1;
MoveCount := Count;
```

```
  end; {MoveCount}

begin
  Rnk := 0;                              {initialize rank}
  for i := 1 to N do
   begin
     Moves := MoveCount(i);              {calculate number of moves i has made}
     if (Rnk mod 2 = 1) then
       remainder := Moves               {add them from left if previous rank odd}
     else
       remainder := i - 1 - Moves;      {add them from right if previous rank even}
     Rnk := i * Rnk + remainder;        {calculate new rank}
   end; {for}
end; {RankPerm}

procedure GetFirstPerm (var Pi : Permutation; var Done : boolean);
 var
  i : integer;
begin
 with Pi do
 begin
  for i := 1 to N + 1 do
   begin                                {Initialize permutation, inverse and direction}
    Value[i] := i;
    Inverse[i] := i;
    Direction[i] := -1;
   end; {for}
  Value[0] := N + 1;
  Activity := [2..N];                   {Initialize the active set}
 end; {with}
 Done := false;
end; {GetFirstPerm}

procedure GetNextPerm (var Pi : Permutation; var Done : boolean);
 var
  j, m : integer;

 function LargestActive : integer;      {returns largest integer in Activity set}
  var
   i : integer;
 begin
  i := N;
  with Pi do
   while not (i in Activity) do
    i := i - 1;
  LargestActive := i;
 end; {LargestActive}

 begin
  with Pi do
```

```
if (Activity <> []) then                              {Activity empty when no more in list}
  begin
  Done := false;                                      {There is another permutation}
  m := LargestActive;                                 {m is value which will move}
  j := Inverse[m];                                    {j is its position}
  Value[j] := Value[j + Direction[m]];                {transpose m with value in direction given by
                                                       Direction}

  Value[j + Direction[m]] := m;
  Inverse[m] := Inverse[m] + Direction[m];            {also transpose position of m with adjacent one}
  Inverse[Value[j]] := j;
  if (m < Value[j + 2 * Direction[m]]) then           {has m reached a boundary?}
    begin
    Direction[m] := -Direction[m];                    {if so, reverse its direction}
    Activity := Activity - [m];                       {and make it passive}
    end; {if then}
  Activity := Activity + [m + 1..N];                  {if not, make all numbers>m active}
  end {if then}
else
  Done := true;
end; {GetNextPerm}
```

A.2 Subsets

Note that a K-subset of an N-set is a K-tuple. Both N and K are global. The binomial coefficients needed for ranking and unranking are global and are assumed to have been computed.

```
const
MaxSetSize = 10;

type
Subset = array[0..MaxSetSize] of integer;
Matrix = array[0..MaxSetSize, 0..MaxSetSize] of integer;

var
N, K : integer;                                       {k-subsets of n-set}
BinCoef : Matrix;                                     {binomial coefficients}

procedure RankSubset (A : Subset; var Rnk : integer);
var
 i : integer;
begin
Rnk := 0;
for i := 1 to K do
 Rnk := Rnk + BinCoef[A[i] - 1, i];
end; {RankSubset}
```

```
procedure UnrankSubset (Rnk : integer; var A : Subset);
 var
  p, m, i : integer;

begin
 m := Rnk;
 for i := K downto 1 do
  begin
  p := i - 1;
  repeat                              {find largest binomial coefficient less than m}
   p := p + 1
  until (BinCoef[p, i] > m);
  m := m - BinCoef[p - 1, i];         {reduce rank by that binomial coefficient}
  A[i] := p;                          {the parameters in the binomial coefficient give the set
                                       value}

  end; {for}
end; {UnrankSubset}

procedure GetFirstSubset (var A : Subset; var Done : boolean);
 var
  i : integer;
begin
 for i := 1 to K do                   {first subset is 1 2 ...}
  A[i] := i;
 A[K + 1] := N + 1;
 Done := false;
end; {GetFirstSubset}

procedure GetNextSubset (var A : Subset; var Done : boolean);
 var
  i, j : integer;
begin
 if (A[1] < N - K + 1) then           {when A[1] is too big, last set has been reached}
  begin
  Done := false;
  j := 0;
  repeat                              {find smallest element that can be advanced }
   j := j + 1
  until (A[j + 1] > A[j] + 1);
  A[j] := A[j] + 1;                   {advance it}
  for i := 1 to j - 1 do              {reset all smaller elements}
   A[i] := i;
  end {if then}
 else
  Done := true;
end; {GetNextSubset}
```

A.3 Set Partitions

The main data type consists of a restricted growth function and a vector whose i^{th} component is the largest entry the restricted growth function may attain in the ith position. The size of the set partitioned is a global variable. The matrix D used in ranking and unranking is global and has been computed elsewhere.

```
const
 MaxSetSize = 12;

type
 Vector = array[0..MaxSetSize] of integer;
 Partition = record
  Value, Maximum : Vector;
  end; {Partition}
 Matrix = array[0..MaxSetSize, 0..MaxSetSize] of longint;   {longint is not standard Pascal}

var
 N : integer;
 D : Matrix;

procedure UnrankSetPart (Rnk : integer; var p : Partition);
 var
  i : integer;

begin
 p.Maximum[1] := 1;                        {start maximum at 1}
 p.Value[1] := 1;                          {first value in RG function}
 for i := 2 to N do
  if (p.Maximum[i - 1] * D[N - i, p.Maximum[i - 1]] <= Rnk) then   {do too many D's fit into Rnk?}
  begin
  p.Value[i] := p.Maximum[i - 1] + 1;      {if too many, make Value as large as possible}
  Rnk := Rnk - p.Maximum[i - 1] * D[N - i, p.Maximum[i - 1]];    {decrease rank}
  p.Maximum[i] := p.Value[i];              {increase max by one}
  end {if then}
  else
  begin
  p.Value[i] := Rnk div (D[N - i, p.Maximum[i - 1]]) + 1;   {if not too many, put them into Value}
  Rnk := Rnk mod (D[N - i, p.Maximum[i - 1]]);    {decrease rank}
  p.Maximum[i] := p.Maximum[i - 1];        {max stays same}
  end; {else}
end; {UnrankSetPart}

procedure RankSetPart (Pi : Partition; var Rnk : integer);
 var
  i, j : integer;
  v, u : Vector;

begin
 Pi.Maximum[1] := 1;                       {Set first max}
```

```
for i := 2 to N do                        {compute Maximum}
  if (Pi.Maximum[i - 1] > Pi.Value[i - 1]) then
    Pi.Maximum[i] := Pi.Maximum[i - 1]    {Value < Maximum means Maximum unchanged}
  else
    Pi.Maximum[i] := Pi.Value[i - 1];     {Value <= Maximum means Maximum increases by
                                          1}
Rnk := 0;                                 {start rank at 0}
for i := N downto 1 do
  Rnk := Rnk + D[N - i, Pi.Maximum[i]] * (Pi.Value[i] - 1);   {use Maximum as index into D table}
end; {RankSetPart}


procedure GetFirstPart (var Pi : Partition; var Done : boolean);
  var
    i : integer;
begin
  Done := false;
  for i := 1 to N do
    with Pi do
      begin
        Value[i] := 1;                    {first RG function is all 1's}
        Maximum[i] := 2;                  {so first Maximum is all 2's}
      end; {with}                         {note: Maximum[1]=2 causes the stopping condition}
end; {GetFirstPart}


procedure GetNextPart (var Pi : Partition; var Done : boolean);
  var
    i, j : integer;

begin
  with Pi do
    begin
      j := N + 1;
      repeat
        j := j - 1
      until (Value[j] <> Maximum[j]);     {find largest non-max component}
      if (j > 1) then                     {if j=1, no more partitions}
        begin
          Done := false;
          Value[j] := Value[j] + 1;       {advance jth component}
          for i := j + 1 to N do
            begin
              Value[i] := 1;              {reset to min past j}
              if (Value[j] = Maximum[j]) then   {is jth component is at its max?}
                Maximum[i] := Maximum[j] + 1    {yes-max past j is one more}
              else
                Maximum[i] := Maximum[j];       {no-max past j is same as max at j}
            end; {for}
        end {if then}
      else
```

```
    Done := true;
  end; {with}
end; {GetNextPart}
```

A.4 Integer Partitions

The data type representing an integer partition is a record which holds the distinct part sizes, the number of each, and the number of distinct parts. The integer partitioned is a global variable. The rank and unrank algorithms are not implemented. They are exercises in Chapter 1.

```
const
MaxInteger = 15;

type
vector = array[0..MaxInteger] of integer;
Partition = record
  Part, Multiplicity : vector;
  NumberOfParts : integer;
  end; {Partition}

var
N : integer;

procedure GetFirstPart (var mu : Partition; var Done : boolean);
begin
with mu do
  begin
  Part[1] := N;                              {first partition has one part of size N}
  Multiplicity[1] := 1;
  NumberOfParts := 1;
  end; {with}
Done := false;
end; {GetFirstPart}

procedure GetNextPart (var mu : Partition; var Done : boolean);
var
k, k1, s, u, v, w : integer;
begin
with mu do
  if (Part[NumberOfParts] > 1) or (NumberOfParts > 1) then
  begin
  Done := false;
  if (Part[NumberOfParts] = 1) then
    begin                                    {smallest partsize is 1}
    s := Part[NumberOfParts - 1] + Multiplicity[NumberOfParts];   {split 1's and next larger}
    k := NumberOfParts - 1;                  {index of part to reduce by one and divide into s}
    end {if then}
  else
```

```
begin                                    {smallest partsize is not 1}
 s := Part[NumberOfParts];               {split last part}
 k := NumberOfParts;                     {reduce this part by one and divide into s}
 end; {else}
w := Part[k] - 1;                        {reduce part by one to divide into s}
u := s div w;                            {this will be the multiplicity of new partsize}
v := s mod w;                            {this will be number of leftover 1's}
Multiplicity[k] := Multiplicity[k] - 1;  {reduce number of these parts}
if (Multiplicity[k] = 0) then
 k1 := k                                 {if none left, make changes at this component}
 else
 k1 := k + 1;                            {if some left, make changes in next component}
Multiplicity[k1] := u;                   {set multiplicity}
Part[k1] := w;                           {set part size}
if (v = 0) then
 NumberOfParts := k1                     {no 1's}
 else
 begin
 Multiplicity[k1 + 1] := 1;              {create block of 1's}
 Part[k1 + 1] := v;
 NumberOfParts := k1 + 1;
 end; {else}
end {if then}
else
Done := true;
end; {GetNextPart}
```

A.5 Product Spaces

An n-tuple in a product space also includes the direction vector and the activity set. Global variables include n and the size of each set in the product space.

```
const
MaximumComponent = 15;

type
vector = array[0.. MaximumComponent] of integer;
IndexSet = set of 0.. MaximumComponent;
Ntuple = record
 Value, Direction : vector;
 Activity : IndexSet;
 end; {Ntuple}

var
N : integer;
MaximumVector : vector;                  {the maximum value at each coordinate}
Base : IndexSet;                         {this set is a list of the non-zero coords of
                                          MaximumVector}
```

```
procedure UnrankProdSpace (Rnk : integer; var v : Ntuple);
var
 i, PrevRnk : integer;
begin
PrevRnk := Rnk;
for i := N downto 1 do
 begin
  v.Value[i] := PrevRnk mod MaximumVector[i];     {start with Value as count from 0}
  PrevRnk := PrevRnk div MaximumVector[i];        {calculate previous rank}
  if (PrevRnk mod 2 = 1) then                      {if previous rank odd count from top}
  v.Value[i] := MaximumVector[i] - v.Value[i] - 1;
 end; {for}
end; {UnrankProdSpace}

procedure RankProdSpace (w : Ntuple; var Rnk : integer);
var
 i, Count : integer;
begin
Rnk := 0;
for i := 1 to N do
 begin
  if (Rnk mod 2 = 1) then                          {if odd, read from top; even, from bottom}
  Count := MaximumVector[i] - w.Value[i] - 1
  else
  Count := w.Value[i];
  Rnk := MaximumVector[i] * Rnk + Count;   {calculate rank from previous rank and count}
 end; {for}
end; {RankProdSpace}

procedure GetFirstVector (var w : Ntuple; var Done : boolean);
var
 i : integer;
begin
with w do
 begin
  for i := 1 to N do
   begin
    Value[i] := 0;                                 {first vector is all 0}
    Direction[i] := 1;                             {all values going up}
   end; {for}
  Activity := Base;                                {all components with non-zero max are active}
 end; {with}
Done := false;
end; {GetFirstVector}

procedure GetNextVector (var w : Ntuple; var Done : boolean);
var
 p : integer;
```

```
function MaxActive : integer;                {returns largest element of Activity}
 var
  i : integer;
 begin
 i := N;
 with w do
  while not (i in Activity) do
   i := i - 1;
  MaxActive := i;
 end; {MaxActive}

begin
 with w do
  if (Activity <> []) then                   {none active means no more}
  begin
   Done := false;
   p := MaxActive;
   Value[p] := Value[p] + Direction[p];      {move largest active component in appropriate
                                              direction}
   if (Value[p] = MaximumVector[p] - 1) or (Value[p] = 0) then      {hit boundary?}
    begin
     Direction[p] := -Direction[p];          {yes-reverse direction}
     Activity := Activity - [p];             {make p inactive}
    end; {if then}
    Activity := Activity + [p + 1..N] * Base;   {no-make all larger than p with non-zero max active}
   end {if then}
  else
   Done := true;
 end; {GetNextVector}
```

A.6 Match to First Available

Lex order is used instead of colex for generating the subsets. However, the structure of the algorithm (GetFirst and GetNext) is the same. While the level in the boolean algebra is a global variable, the GetFirst subroutine has the subset size as an input parameter because it is called to generate the k-subsets and the (k+1)-subsets. This additional parameter is not necessary in GetNext since GetNext computes the size of the subset from the previous one. An additional global variable is a set which represents those (k+1)-subsets which have been matched. The data type used to represent a subset is different: we use Pascal subsets here.

```
const
 MaxSetSize = 12;
 MaxBinCoef = 1000;

type
 Subset = set of 0..MaxSetSize;
 SubsetList = set of 1..MaxBinCoef;
```

```
var
N, K : integer;
UsedSubsets : SubsetList;

procedure GetFirstSubset (p : integer; var S : Subset; var Done : boolean);
begin
S := [1..p];
Done := false;
end; {GetFirstSubset}

procedure GetNextSubset (var S : Subset; var Done : boolean);
 var
  Count, j, i : integer;
begin
j := N;
 repeat                                    {find largest element that can advance}
 j := j - 1;
 until ((j in S) and not (j + 1 in S)) or (j = 0);
 if (j > 0) then                           {j=0 means no more subsets}
 begin
 Done := false;
 Count := 0;
 for i := j to N do                        {count number in subset past this element}
  if (i in S) then
   Count := Count + 1;
  S := (S - [j..N]) + [j + 1..j + Count];  {remove these and j and add Count contiguous
                                           elements}

 end {if then}
 else
 Done := true
end; {GetNextSubset}

function Matched (A : Subset; var B : Subset) : boolean;      {returns true and subset B if match}
 var
  ListDone, StopLoop : boolean;
  p : integer;
begin
GetFirstSubset(K + 1, B, StopLoop);
p := 1;
 while not StopLoop do                     {search for first match or end of list}
 if (A <= B) and not (p in UsedSubsets) then    {A is subset of B and B is not used}
  StopLoop := true                         {found a match}
 else
 begin
  GetNextSubset(B, ListDone);              {go to next subset}
  StopLoop := ListDone;
  p := p + 1;
 end; {else}
 if not ListDone then                      {ListDone will be true if no match found}
```

```
begin
  Matched := true;                              {found a match}
  UsedSubsets := UsedSubsets + [p];             {make it unavailable}
  end {if then}
else
  Matched := false;                             {no match}
end; {Matched}


procedure MatchToFirst;
var
  A, B : Subset;
  EndOfList : boolean;
  i : integer;
begin
  UsedSubsets := [];                            {start with no subsets used}
  GetFirstSubset(K, A, EndOfList);              {first subset to match}
  while not EndOfList do                        {search list of k-subsets in lex order}
  begin
    if Matched(A, B) then                       {find matching (k+1)-subset, if exists}
      PrintPair(A, B)
    else                                        {no matching subset}
      PrintNoMatch(A);
    GetNextSubset(A, EndOfList);
  end; {while}
  GetFirstSubset(K + 1, B, EndOfList);
  i := 1;
  while not EndOfList do                        {now search list of (k+1)-subsets in lex order}
  begin
    if not (i in UsedSubsets) then              {list unmatched ones}
      PrintNoMatch(B);
    i := i + 1;
    GetNextSubset(B, EndOfList);
  end; {while}
end; {MatchToFirst}
```

A.7 The Schensted Correspondence

We give here the implementation of the Schensted algorithm, both for encoding a permutation as a pair of tableaux and vice versa. A partition is given as a record which contains the number of parts and a vector of parts, largest first. A tableau consists of two partitions (the shape and its conjugate) and a matrix of entries. A pair of tableaux is the record TableauPair with Bumping and Template as its two constituents. Finally, a permutation in two-line notation is kept as the record TwoLinePerm consisting of two vectors, TopRow and BottomRow.

The main work of the algorithm is done in the two procedures SchenstedInsert and SchenstedDelete, which insert a value into a tableau and delete a value from a tableau, respectively.

The number of cells (or the number of elements in the permutation) is the global variable N.

```
const
 MaximumNumParts = 20;

type
 vector = array[0..MaximumNumParts] of integer;
 matrix = array[0..MaximumNumParts] of vector;
 TwoLinePerm = record
  TopRow, BottomRow : vector;
  end;
 Partition = record
  NumberOfParts : integer;
  Part : vector;
  end;
 Tableau = record
  Shape, Conjugate : Partition;
  CellEntry : matrix;                      {entries of tableau, in matrix form}
  end;
 TableauPair = record
  Bumping, Template : Tableau;             {Bumping sometimes called the P-tableau; Template
                                           the Q-tableau}
  end;

var
 N : integer;

procedure SameShape (var S : Tableau; T : Tableau);     {Makes the tableau S the same shape as
                                                        T}
begin
 S.Shape := T.Shape;
 S.Conjugate := T.Conjugate;
end; {SameShape}


{changes the kth component of Lambda by plus or minus 1 (TrimSize)}
{if Shorten is true, changes the number of parts by the same amount}
{this will happen when part k has size 1.}
procedure TrimShape (var Lambda : Partition; k : integer; Shorten : boolean; TrimSize : integer);
begin
 with Lambda do
  begin
  Part[k] := Part[k] - TrimSize;
  if Shorten then
   NumberOfParts := NumberOfParts - TrimSize;
  end; {with}
end; {TrimShape}


procedure DeleteCell (var P : Tableau; Row, Col : integer);     {remove cell from P at Row,
                                                               Col}
begin
 with P do
  begin
```

```
   TrimShape(Shape, Row, (Col = 1), 1);
   TrimShape(Conjugate, Col, (Row = 1), 1);
  end; {with}
 end; {DeleteCell}

{uses Schensted insertion to insert k into tableau P.  Row and Col are new cell added to P}
 procedure SchenstedInsert (var P : Tableau; k : integer; var Row, Col : integer);
  var
  x : integer;
  DoMore : boolean;
 begin
  with P do
  begin
   Col := 1;                          {start at first column}
   DoMore := true;
   while DoMore do
    if (k <= CellEntry[Conjugate.Part[Col], Col]) then        {k bumps something}
    begin
     Row := Conjugate.Part[Col];            {start looking for value to bump}
     repeat                                 {find value to bump}
      Row := Row - 1
     until (CellEntry[Row, Col] < k);
     Row := Row + 1;                        {value to bump is one cell down}
     x := CellEntry[Row, Col];              {x is value to bump}
     CellEntry[Row, Col] := k;              {replace it with k}
     k := x;
     Col := Col + 1;                        {move out to next column}
    end {if then}
    else                                    {k goes at end of column}
    begin
     Row := Conjugate.Part[Col] + 1;        {Row is row index}
     CellEntry[Row, Col] := k;              {place k at end of column}
     DoMore := false;                       {no more bumping}
     TrimShape(Shape, Row, (Col = 1), -1);  {add cell to shape and conjugate}
     TrimShape(Conjugate, Col, (Row = 1), -1);
    end; {else}
  end; {with}
 end; {SchenstedInsert}

{Schensted deletion; value in cell of P at Row, Col starts deletion; x is the value removed}
 procedure SchenstedDelete (var P : Tableau; var x : integer; Row, Col : integer);
  var
  i, j, y : integer;
 begin
  with P do
  begin
   x := CellEntry[Row, Col];              {start with x as value in P at Row,Col}
   DeleteCell(P, Row, Col);               {remove this cell from P}
   for j := Col - 1 downto 1 do           {bumping}
```

```
      begin
      i := 1;
      repeat                                 {find value to bump}
        i := i + 1
      until (CellEntry[i, j] > x) or (i > Conjugate.Part[j]);     {don't go past end of column}
      i := i - 1;                            {value to bump in row above}
      y := CellEntry[i, j];                  {bump it}
      CellEntry[i, j] := x;
      x := y;                                {repeat for next column down}
      end; {for}
    end; {with}
  end; {SchenstedDelete}


{convert permutation pi in two-line form to pair of tableaux}
procedure SchenstedEncode (pi : TwoLinePerm; var Z : TableauPair);
var
  i, Row, Col : integer;

  procedure Empty (var T : Tableau);        {make T the empty tableau}
  var
    i : integer;
  begin
    with T do
    begin
      Shape.NumberOfParts := 0;
      Conjugate.NumberOfParts := 0;
      for i := 1 to MaximumNumParts do
      begin
        CellEntry[i, 0] := 0;
        CellEntry[0, i] := 0;
        Shape.Part[i] := 0;
        Conjugate.Part[i] := 0;
      end; {for}
      CellEntry[0, 0] := 0;
    end; {with}
  end; {Empty}


begin
  Empty(Z.Bumping);                         {start with empty tableaux}
  Empty(Z.Template);
  for i := 1 to N do
  begin
    SchenstedInsert(Z.Bumping, pi.BottomRow[i], Row, Col); {Schensted insert permutation value}
    Z.Template.CellEntry[Row, Col] := pi.TopRow[i]; {insert top row value into new cell}
  end; {for}
  SameShape(Z.Template, Z.Bumping);         {Bumping has right shapes; make Template same}
end; {SchenstedEncode}


{convert pair of tableaux to a permutation in two-line form}
procedure SchenstedDecode (Y : TableauPair; var pi : TwoLinePerm);
```

```
var
 i, k, Row, Col : integer;
 Z : TableauPair;

procedure FindMaximum (Q : Tableau; var Row, Col : integer);      {returns Row and Col of largest
                                                                   member of Q}
 var
  x, j : integer;
begin
 x := 0;
 with Q do
  for j := Conjugate.NumberOfParts downto 1 do
   if (CellEntry[Conjugate.Part[j], j] > x) then
    begin
     Row := Conjugate.Part[j];
     Col := j;
     x := CellEntry[Row, Col];
    end; {if then}
end; {FindMaximum}

function NumberOfCells (T : Tableau) : integer;      {returns number of cells in T}
 var
  i, sum : integer;
begin
 sum := 0;
 with T do
  for i := 1 to Shape.NumberOfParts do
   sum := sum + Shape.Part[i];
 NumberOfCells := sum;
end; {NumberOfCells}

begin
 Z := Y;                                    {Z is going to be clobbered}
 N := NumberOfCells(Z.Bumping);             {compute N}
 for i := N downto 1 do                     {remove each cell of Template}
  begin
   FindMaximum(Z.Template, Row, Col);
   pi.TopRow[i] := Z.Template.CellEntry[Row, Col];   {top row value is maximum in template
                                                      tableau}
   SchenstedDelete(Z.Bumping, k, Row, Col); {Schensted delete from this cell}
   pi.BottomRow[i] := k;                    {value bumped out is bottom row value}
   DeleteCell(Z.Template, Row, Col);        {remove cell from template}
  end; {for}
end; {SchenstedDecode}
```

A.8 The Prüfer Correspondence

The data type for a tree is a vector of subsets. The vth subset is the set of vertices adjacent to vertex v. This is called the adjacency list. The number of vertices, N, is a global variable.

```
const
 MaxTreeSize = 20;

type
 Vertex = 0..MaxTreeSize;
 VertexSet = set of Vertex;
 Tree = array[1..MaxTreeSize] of VertexSet;  {tree is adjacency list}
 Vector = array[1..MaxTreeSize] of Vertex;

var
 N : integer;

procedure DecodeVector (a : Vector; var T : Tree);
 var
 i, v, w : Vertex;
 Degree : Vector;

procedure AddAnEdge (var T : Tree; v, w : Vertex);        {add edge (v,w) to T}
 begin
 T[v] := T[v] + [w];
 T[w] := T[w] + [v];
 end; {AddAnEdge}

function Largest (Degree : Vector) : Vertex;        {returns largest terminal vertex of tree with
                                                      given degrees}
 var
 i : Vertex;
 begin
 i := N;
 while (i >= 1) and (Degree[i] <> 1) do
 i := i - 1;
 Largest := i;
 end; {Largest}

function Smallest (Degree : Vector) : Vertex;        {returns smallest terminal vertex of tree with
                                                       degree given}
 var
 i : Vertex;
 begin
 i := 1;
 while (i <= N) and (Degree[i] <> 1) do
 i := i + 1;
 Smallest := i;
 end; {Smallest}
```

```
begin
for v := 1 to N do
 Degree[v] := 1;
for i := 1 to N - 2 do
 Degree[a[i]] := Degree[a[i]] + 1;              {compute degrees}
for v := 1 to N do
 T[v] := [];                                    {start with empty tree}
for i := 1 to N - 2 do
 begin
 w := Largest(Degree);
  AddAnEdge(T, a[i], w);                         {add edge from largest terminal vertex to vertex a[i]}
 Degree[a[i]] := Degree[a[i]] - 1;              {reduce degree}
  Degree[w] := 0;                                {kill largest terminal vertex}
 end; {for}
v := Largest(Degree);                           {two vertices remain; connect them}
w := Smallest(Degree);
AddAnEdge(T, v, w);
end; {DecodeVector}

procedure EncodeTree (T : Tree; var a : Vector);
var
 k, v : Vertex;

 function Largest (T : Tree) : Vertex;          {returns largest terminal vertex of tree T}
 var
  i : Vertex;

  function SetSize (S : VertexSet) : Vertex;    {returns size of set S}
  var
  m, i : Vertex;
  begin
  m := 0;
  for i := 1 to N do
   if (i in S) then
    m := m + 1;
  SetSize := m;
  end; {SetSize}

 begin
 i := N;
 while (i >= 1) and (SetSize(T[i]) <> 1) do     {set size is 1 when vertex is terminal}
  i := i - 1;
 Largest := i;
 end; {Largest}

 function Adjacent (T : Tree; v : Vertex) : Vertex;   {returns vertex adjacent to v}
 var
  k : Vertex;
 begin
 k := 1;
```

```
  while (k <= N) and (not (k in T[v])) do
  k := k + 1;
  Adjacent := k;
  end; {Adjacent}

procedure RemoveEdge (var T : Tree; v, w : Vertex);        {remove edge (v,w) from T}
  begin
  T[v] := T[v] - [w];
  T[w] := T[w] - [v];
  end; {RemoveEdge}

begin
  for k := 1 to N - 2 do
   begin
    v := Largest(T);                          {get largest terminal vertex}
    a[k] := Adjacent(T, v);                   {set a[k] to vertex adjacent to largest terminal vertex}
    RemoveEdge(T, v, a[k]);                   {remove this edge}
   end;
  end; {EncodeTree}
```

A.9 The Involution Principle

We give here only the involution-principle. We omit the definitions of the two involutions ϕ and ψ. The underlying set is of type YourDataType. For example, if we were to use this procedure to implement the involutions which proves Euler's theorem (§4.6), YourDataType would be pairs of partitions: one with only distinct parts of even size and one with no restrictions.

Pascal imposes some restriction on the way the involution principle can be programmed. Because Pascal **functions** have only a narrow choice of data types as output type, the involutions ϕ and ψ must, in general, be represented as **procedures** with input and output parameters.

```
function AreEqual(P, Q : YourDataType) : boolean;
  begin
  {return true if P=Q}
  {return false if P≠Q}
  end; {AreEqual}

procedure Phi (P : YourDataType; var Q : YourDataType);
  begin
  {Q=Phi(P)}
  end; {Phi}

procedure Psi (P : YourDataType; var Q : YourDataType);
  begin
  {Q:=Psi(P)}
  end; {Psi}
```

```
{input is Lambda; output is Rho; IsFixedPt is false if Lambda is not a fixed point}
procedure InvolPrin (Lambda : YourDataType; var Rho : YourDataType; var IsFixedPt :
                                            boolean);
var
 W, X, Y, Z : YourDataType;

begin
X := Lambda;
Phi(X, Y);
if AreEqual(X, Y) then                        {X is in fixed point set of phi}
 begin
  repeat
   Z := Y;
   Psi(Z, W);                                 {apply Psi followed by Phi}
   Phi(W, Y);
  until AreEqual(Z, W) or AreEqual(W, Y);   {until a fixed point reached}
  Rho := W;
  IsFixedPt := true;
 end {if then}
else
 begin
 Psi(X, Y);
  if AreEqual(X, Y) then                      {X is in fixed point set of psi}
  begin
   repeat
   Z := Y;
    Phi(Z, W);                                {apply Phi followed by Psi}
    Psi(W, Y);
   until AreEqual(Z, W) or AreEqual(W, Y);      {until a fixed point reached}
   Rho := W;
   IsFixedPt := true;
  end {if then}
  else
   IsFixedPt := false;                        {input not fixed point of Phi or Psi}
 end; {else}
end; {InvolPrin}
```

Index

Antichain, 32

Ballot problem, 60
Bell number, 18
Bijection, 7, 57
    signed, 141
    weight-preserving, 58
Binomial
    coefficient, 7
    theorem, 7
Bit string, 63
Block, 18
Boolean algebra, 28, 33
Broken circuit, 155
Brother, 59

Canonical cycle decomposition, 75
Catalan numbers, 59
Cayley's theorem, 64
Cayley-Hamilton theorem, 120
Chain, 33
    decomposition, 33
    decomposition, symmetric, 40
    maximal, 52
    product of, 30
Chromatic polynomial, 154
Colex order, 8
Colexicographic order, 8
Column
    deletion, 90
    insertion, 86
    strict, 81
Component, 59
Conjugate, 12
Content, 81
Cover, 26
    relation, 26
Cut-edge, 69

Cycle
    of a graph, 58
    notation, 7
    of a permutation, 7

Decomposition number, 103
Degree, 58
Descent, 76
Dictionary order, 8
Digraph, 58
    functional, 105
Directed
    cycle, 58
    edge, 58
    graph, 58
Dominance, 29
Durfee square, 72

Edge, 58
    directed, 58
Erdős-Ko-Rado theorem, 45
Erdős-Szekeres theorem, 108
Euler's pentagonal number theorem, 111
Euler's theorem, 70
Eulerian number, 77

Fall, 76
Falling factorial, 2
Father, 59
Ferrers diagram, 12
Filament, 103
Fixed point set, 110
Forest, 69
Frobenius formula, 130

Gale-Ryser theorem, 108
Generating function, 57

Graph, 58
    bipartite, 59
    complete, 59
    component of a, 59
    connected, 58
    cycle of a, 58
    directed, 58
    labeled, 59
    multi-, 122
    proper coloring of a, 154
    simple, 58
    strongly connected, 59
    unlabeled, 59
Gray code, 15
Greatest lower bound, 32

Hall's theorem, 50
Hamiltonian cycle, 15
Hasse diagram, 27
Hook, 84
    formula, 84

Incidence matrix, 51
Inclusion-exclusion principle, 110
Independent set, 32
Index, 107
Interlace, 53
Inverse, 3
Inversion
    lagrange, 105
    of a permutation, 31
    poset, 27
    sequence, 7
Involution, 98, 110
    principle, 141, 178
    sign-reversing, 110
    weight-preserving, 122

Jacobi triple product, 112
Jacobi-Trudi identity, 147
Johnson-Trotter algorithm, 2
Join, 32

Kostka number, 83
Kruskal-Katona theorem, 45

Lagrange inversion, 105

Lattice, 32
    domination, 29
    partition, 28
    path, 36, 130
    permutation, 108
    Young's, 29
Least upper bound, 32
Level set, 31
Lex order, 2
Lexicographic order, 2, 8
    reverse, 12
Linear extension, 50
Linear order, 1
Listing algorithm, 1
Littlewood-Offord problem, 39
Log-concave, 52
Loop, 58

Matching, 33
Matrix-tree theorem, 125
Maximum, 32
Meet, 32
Minimum, 32
Monotone, 102
Multi-graph, 122
Multinomial
    coefficients, 80
    theorem, 80
Multiple edge, 58
Multiplicity, 11, 80
Multiset, 40, 79
    permutation, 80

Order symmetric, 31
Orthogonality, 77

Part, 11
Partially ordered set, 26
Partition
    block of a, 18
    conjugate, 12
    integer, 11, 69, 166
    lattice, 28
    part of a, 11
    self-conjugate, 72
    set, 18, 164
Pascal's triangle, 8

Path, 58
    directed, 59
    simple, 58
Pentagonal number, 111
Permutation, 2, 73, 159
    cycle of a, 7
    in cycle notation, 7
    descent of a, 76
    even, 22
    fall of a, 76
    lattice, 108
    multiset, 80
    in one-line notation, 7
    product of, 6
    run of a, 76
    sign of a, 110
    in two-line notation, 7
Polya's theorem, 59, 81
Poset, 26
    inversion, 27
    maximum of a, 32
    minimum of a, 32
Principle of inclusion-exclusion, 110
Product
    of chains, 30
    of permutations, 6
    space, 15, 167
Proper coloring, 154
Property, 110
Prüfer correspondence, 64, 176

q-binomial coefficient, 72

Rank
    in a list, 1
    of a poset, 31
    symmetric, 31
    unimodal, 31
Ranking algorithm, 1
Refinement, 28
Reflection principle, 130
Restricted growth function, 18
RG function, 18
Rogers-Ramanujan identities, 143
Root, 59
Row insertion, 109
Run, 76

Schensted correspondence, 85, 93, 171
Schur function, 139
Score vector, 116
Shape, 62, 81
Sign
    of a permutation, 110
    -reversing, 110
Signed set, 110
Simon Newcomb's problem, 107
Son, 59
    left, 59
    right, 59
Sperner property, 32
Sperner's theorem, 32
Stirling number
    of the first kind, 77
    of the second kind, 18
Subsequence, 93
Successor algorithm, 9
Symmetric function
    complete, 137
    homogeneous, 137

Tableau, 81
    column strict, 81
    content of a, 81
    shape of a, 81
    standard, 60, 83
Tournament, 115
    non-transitive, 116
    transitive, 115
Transposition, 6
    adjacent, 6
Tree, 59
    binary, 59
    full binary, 60
    labeled, 64
    ordered, 59
    root of a, 59
    rooted, 59
    spanning, 125

Unimodal, 31
Unrank, 1

Vandermonde's determinant, 114
Vandermonde's theorem, 25

Vertex, 58
    adjacent, 58
    internal, 61
    terminal, 61

Weight, 78, 115
    -preserving, 58, 78
Well-formed parentheses, 60
Whitney number, 51
Winner, 115

Young diagram, 12
Young's lattice, 29

Undergraduate Texts in Mathematics

continued from ii

Martin: The Foundations of Geometry and the Non-Euclidean Plane.

Martin: Transformation Geometry: An Introduction to Symmetry.

Millman/Parker: Geometry: A Metric Approach with Models.

Owen: A First Course in the Mathematical Foundations of Thermodynamics.

Prenowitz/Jantosciak: Join Geometrics.

Priestly: Calculus: An Historical Approach.

Protter/Morrey: A First Course in Real Analysis.

Protter/Morrey: Intermediate Calculus.

Ross: Elementary Analysis: The Theory of Calculus.

Scharlau/Opolka: From Fermat to Minkowski.

Sigler: Algebra.

Simmonds: A Brief on Tensor Analysis.

Singer/Thorpe: Lecture Notes on Elementary Topology and Geometry.

Smith: Linear Algebra. Second edition.

Smith: Primer of Modern Analysis.

Thorpe: Elementary Topics in Differential Geometry.

Troutman: Variational Calculus with Elementary Convexity.

Wilson: Much Ado About Calculus.